Combat Vet Don't Mean Crazy:
Veteran Mental Health in Post-Military Life

Duane K. L. France, MA, MBA, LPC

COMBAT VET DON'T MEAN CRAZY.

Published By:
NCO Historical Society
P.O. Box 1341
Temple, TX 76503
www.ncohistory.com

 NCO Historical Society

The content of this book has been previously published in a digital format on the Head Space and Timing blog, located at www.veteranmentalhealth.com

Cover Design by Extended Imagery

The author of this book is a Mental Health Counselor licensed to practice in the state of Colorado. The thoughts, ideas, musings, and posts in this book come from his military experience, professional experience, and personal opinions. They do NOT, however, represent professional advice. While he is a Mental Health Counselor, he is not YOUR Mental Health Counselor, and the guidance in this work should not be considered a substitute for working with a licensed clinical mental health provider. The opinions expressed here are his own, and in no way should be seen as reflection of his agency, his profession, or any professional associations that he is connected with.

ISBN: 0-9963181-4-3

ISBN-13: 978-0-9963181-4-3

Published in the United States of America

1st Edition

To My Family: Connie, Christina, and Daniel. You served along with me, in a different way, but with the same pride and sacrifice as I did

To My Family: The Soldiers I served with in all of my units, but especially the troops of A Co, 704th BSB and those that rode with us. Having been through the valley of death, I would do it all again if you were by my side.

And finally, To My Family: all who have stepped up to serve our nation through military service. I don't know you, but you're my brother and sister. You're not alone...ever.

CONTENTS

Foreword

Duane France provides a tribute to veterans in *Combat Vet Don't Mean Crazy: Veteran Mental Health in Post-Military Life*. This body of work is a collection of rich stories that depict the horrors of war and the transition back to civilian life. Each section paints an all sensory picture of the internal struggles on the back-drop of negative societal stigma and stereotypes. The book confirms how these societal issues impact the mental and physical well-being of veterans and their family members. In this book, the reader is provided with a better understanding of the wounded warrior's moral injury that occurs in some veterans after combat operations. As a consequence, thousands of veterans have become homeless, unemployed, and struggle to claw their way back to some level of normalcy and balance within civilian life. Indeed, the combination of mental and physical disabling conditions post-deployment and during reintegration imprints on the veteran's mind, body, and spirit.

Duane's extensive military career as an Army NCO with five combat and operational deployments in Iraq, Afghanistan, North Africa, and Bosnia-Herzegovina has laid a strong foundation for his body of work. Duane's background as a veteran and as a mental health professional provides a unique prospective in working with the military culture. Duane comprehensively communicates, from the veteran's point of view, critical elements of life after the military. He draws on a range of other behavioral health experiences which includes his roles as Director of Veteran Services Family Care Center and Executive Director for the Colorado Veterans Health and Wellness Agency in Colorado Springs. Duane continues to enhance the quality of life for veteran services as a mentor, advocate, and an innovated service provider for veterans and their family members. Thus, Duane's body of work bridges the gap between real-life experiences of veterans and limitations found in the traditional research literature.

Duane's compassion and heart for veteran-related issues are further exemplified by the podcast he created for the Change Your POV Podcast Network titled Head Space and Timing. Here, in his weekly podcast, he explores and defines the military and veteran's way of life in a manner rarely seen. He presents first-hand accounts of service members and veterans' experiences in the deployment cycle through transition into civilian life. This adds immense value in *Combat Vet Don't Mean Crazy* as a qualitative study of sorts.

His unique perspectives in the chapter, *"Cross-Generational War"* suggests that for the first time in American history, that a child born around 9/11 could potentially be deployed OCONUS and fight alongside his or her father or mother. Indeed, there is a story to tell that should inspire readers to research the inter-generational implications of how service members and their family members transition well, in-and-out of military service. In the piece, *"You Break it You Buy it"* Duane points to critical issues that some veterans experience as feeling like they are "damaged goods" or "broken". These statements bring to light the overwhelming number of thoughts, feelings, and experiences that occur within veterans on a day-to-day basis. This work speaks volumes to the psychological impact resulting from repeated exposure to combat operations.

One of the many gems of this body of work is found in chapter *"Six Thoughts on Posttraumatic Stress Disorder"*. This section speaks to the fact that PTSD is not just a psychological condition. Rather, it is a whole-body experience that has physiological, neurobiological, and cognitive-relational significance. Duane reports on some current thinking in this specialty area of mental health suggesting that PTSD is not a single-episode event. In military service, there are multiple critical incidents which are cumulative and are seated within the veterans' mind, body, and spirit. Duane attempts to destigmatize the "D" in PTSD to emphasize the human nature of the natural body's reaction to extraordinary stressful and traumatic events. The clinical diagnostic category does not quite address the interaction between the veteran and his/her environment. Nor does it provide truth to the fact that warfighters are human beings. The consequence is that service members, during combat operations, are trained to adapt and survive. However, it is not until later in post-deployment and reintegration that the veteran is absent of his/her identity of a Hollywood-style super-hero.

Overall, the themes communicated in Duane's work addresses how some veterans thrive and bounce-back from adversity and cultivate resiliency after exposure to the human pain and suffering of war. The path home for the war-fighter is a long journey for some. Bringing meaning and purpose to ones' military experiences into the present are eloquently communicated in *Combat Vet Don't Mean Crazy*. It is living in-the-present moment that many veterans find transformation for optimal living in balance and with good mental, physical, and spiritual well being.

Mark A. Stebnicki, Ph.D., LPC, DCMHS, CRC, CCM, CMCC
Professor, Coordinator of Military and Trauma Counseling Certificate Program
Department of Addictions and Rehabilitation
East Carolina University

PART 1

Raising Awareness about the Psychological Impact of Military Service

The problem with the stigma around mental health is really about the stories that we tell ourselves as a society. What is normal? That's just a story that we tell ourselves. - Matthew Quick

The first step towards making any change in our lives is to become aware that change is necessary. That awareness can come by many paths: self-reflection, crisis, others pointing it out to us. In order for the conversation about veteran mental health to change, there first needs to be an awareness about how service in the military changes someone.

When a civilian joins the military, they experience an intense, isolating experience that is designed to assimilate them into the military culture. They learn discipline, highly technical and tactical skills, and the need to work as a team to accomplish a difficult task. Service in the military changes the way that someone thinks and acts. Whether they have served five years or twenty-five years, the impact of the military service is one that develops a connection to an entirely different culture.

By any definition of the word, the military is a different culture. Merriam-Webster defines culture in the following ways:

1. The characteristic features of everyday existence shared by people in place or time

2. The set of shared attitudes, values, goals, and practices that characterizes an institution

3. The set of values, conventions, or social practices associated with a particular field, activity or societal characteristic

4. The integrated pattern of human knowledge, belief, and behavior that depends upon the capacity for learning and transmitting knowledge to succeeding generations

Each of these has application to military service. The characteristic features of everyday existence: the hierarchical structure of military rank, duty positions, training events. The set of shared attitudes, values, goals, and practices: the Army Values. The Rifleman's Creed. The Warrior Ethos. The set of values, conventions, or social practices associated with a particular field: military ceremonies. Traditions. The integrated pattern of human knowledge transmitted to succeeding generations: military history.

Along with this culture is the thoughts, attitudes, and mindset of the culture. A veteran from one country has an affinity for a veteran from another country, sometimes finding more in common with another veteran than they do with their neighbors who haven't served.

The following collection of articles is an attempt to help raise the awareness about the psychological impact of military service. They try to convey the struggle that service members have when leaving the military and rejoin their community, the barriers that exist in getting the support they need, and discussions on prevalent attitudes towards suicide and posttraumatic stress disorder. By raising community awareness, we make it easier for veterans to be open and honest about how their military service has changed the way the look at and think about the world.

Starving at the Feast: A Parable of Modern Transition

The traveler stepped through the gate, excited and anxious about the future. Both the possibilities and challenges seemed endless, and he was curious what support he could obtain to help him along his journey. He clutched in his hand his traveling papers.

As he walked into the village clustered outside the gate of the garrison, the first thing that surprised the traveler was the crowded square, a wide space of clamoring chaos that had an overwhelming number of vendors, all vying for his attention. The selection of choices seemed endless, as well as the variety; there were vendors that seemed to want to give him things, and others that wanted to take things from him.

He felt a tug on his garment; he turned to see an older gentleman, kind; he was perhaps a traveler in his youth. "A new companion on the path, I see," the gentleman said with an honest smile. "You'll want to register with the Traveler's Agency over there, if they'll have you." The traveler looked to his left, and a large imposing structure loomed. A line of travelers, both young and old, stretched out the door and around the block. "No, I think I'm okay," the traveler said, "I want to get started and don't think I need that kind of assistance. Besides, it looks like those other travelers need it more than me."

The kind gentleman's face fell slightly. "Don't wait too long, son. You could end up walking a rough road for a while, like me." The traveler thanked him and continued into the square.

The assortment of vendors continued to startle the traveler. Where to begin? There were some obvious well-established and long-serving vendors, lined together on a high rise on the far side of the square. Their stalls were strong and well-made, but it was curious…it did not seem that many travelers were interested in them. The representatives of these vendors, grave and severe in their demeanor, could be seen in a serious discussion with the King's ministers, but travelers passed to and fro in front of their establishments without giving them a second glance.

Perhaps more promising were the newer vendors on the left side of the square, no less grand than the established vendors, but with a different air about them. A sense of eagerness, of youth, and perhaps even a touch of new wealth drove their activities. On closer inspection, however, it was hard to determine which of these newer vendors were here to support the traveler, or were here to take advantage of the traveler. The lack of distinction caused the traveler to pause.

On his right, the traveler could see the specialty vendors, who promised all manner of support. He could see vendors that promised to help find employment, others that promised housing, a third that was distributing professional garments. He could see a large number of vendors promising a life of adventure; a week-long trip to the King's Forest, a weekend adventure on Queen's Lake. The traveler looked at these offers longingly, recognizing them as trips of a lifetime, but failed to see how they could help him along his journey.

Interspersed among all of these vendors were others, operating small stalls in an inefficient way. Why was there so much duplication in their efforts, the traveler wondered? Every third person wanted to help the traveler record his experiences so that he may find employment. The conflicting voices clamoring in his ears was beginning to overwhelm…

Adding to this cacophony, and confusion, were the minstrels and the traveling performers. Each individual and group with similar advice, raising their voices to try to be heard above the other, because the traveler *needed* to hear what they had to say, in order to be successful on his journey! A vast array of information was starting to wash over the traveler, from all sides.

He took a deep breath and decided that he should start somewhere. He stepped toward one of the stalls promising employment support. "I'd like to find work," the traveler said, "but I also could use help finding lodgings."

"Let me see your traveling papers, sir" the kind young woman said. After glancing over his documents, she said, "Yes sir, you have the qualifications for us to support you. We don't offer support for lodging, however, so you must seek that elsewhere."

The traveler was confused. "You don't support all travelers?" He said. "No," came the reply, "Only recent travelers with the right kind of papers."

"And you can't help me secure lodging?"

"No," came the reply again, "We only support employment."

The traveler thanked her and moved on, past another vendor offering an adventure. As appealing a diversion as that would be, he recognized that it was even less beneficial than it was before. He hurried towards a vendor that promised support with lodging. "I'd like to get some help finding somewhere to stay," the traveler said. Again he was asked, "Could I see your papers?" After a brief scan, they were returned to him. "We can help provide lodging for a short time, but you will need employment before we can support you." At this point, frustration began to set in. "Could you not help me find employment as well?" The traveler asked. "No, I'm sorry, I'm afraid not. I can only help with lodging."

As he stepped away from the stall, he began to wonder…if I'm having such a hard time, what about those who are not new travelers? Or who do not have the right paper? He looked down the lane, and saw another traveler being detained by the Sheriff's men. Could that be me, he thought?

The traveler stood in the middle of the square, overwhelmed by choice and frustrated by obstacles, and began to starve in the middle of a vast feast.

One of the Most Important Questions about Veteran Mental Health is Who's Paying the Bill

There are several challenges regarding veteran mental health, and the combination of them can often make the task of accessing mental health services impossible for veterans. There is, of course, the stigma attached to seeking mental health services. There is also the need for counselors and therapists to be culturally competent, both because they need to be aware of what their client is going through, but also so that the veteran is able to connect with and trust their mental health provider.

An additional challenge: who's going to pay for it. I am personally of the opinion that veterans should not have to pay for their own mental health counseling. I recognize that there is debate about this: the client should have some buy-in through co-pays and no-show fees. Without having some skin in the game it is easier for the veteran to not show up for a scheduled appointment, and I understand all of that. From my perspective, however, veterans *have* paid for it, through their service and their experiences. I do have some flexibility, of course; the insurance copays for an outpatient mental health counseling session is less than $40. If a veteran is not willing to pay that for something that can benefit them, then a conversation about commitment probably has to happen.

I work with veterans involved in the criminal justice system. Once a veteran has pleaded guilty to a charge, we have a contract with the local probation department to be paid for the services we provide. Once a veteran is in the program, we can provide weekly individual and group counseling, and more if needed. My problem with this is, why does a veteran have to first commit a crime, and then plead guilty to that crime in order to have full access to services? Wouldn't it be better if the veteran were to receive treatment *first*, in order to avoid involvement in the courts?

There are several attempts to solve these challenges. Masters level mental health counselors are now eligible to participate in the Department of Veterans Affairs Choice program, after a change that happened in December of 2014 and a contract change in the spring of 2015[1]. Originally, Licensed Professional Counselors (LPCs) and Licensed Marriage and Family Therapists (LMFTs) were excluded from participating in the program because the original guidance required that all Choice providers be eligible to participate in Medicare.

This is the crux of the challenge: unfortunately, mental health professionals with licensure of LPCs and LMFTs are currently prohibited from being reimbursed under Medicare[2]. As it stands, the only Masters-level mental health counselors eligible for reimbursement are Licensed Clinical Social

Workers (LCSW), even though LPCs and LMFTs are recognized at the state level to be able to perform the same services.

The fact that LPCs are ineligible for reimbursement under Medicare is significant for several reasons. Federal agencies such as the Department of Defense and the Department of Veterans Affairs often require that mental health practitioners be required to be eligible for Medicare reimbursement, regardless of state licensure. By allowing LPCs to be included in this category, the overwhelming need for mental health professionals in these two agencies can be addressed.

Another challenge is that I often see veterans who are medically retired and have both Medicare and TRICARE. TRICARE requires that other health insurance be billed before TRICARE is billed; therefore, a veteran seeking services at our agency, who was medically retired due to a service-connected disability, is essentially barred from receiving services because the majority of our clinicians are not eligible to be reimbursed by Medicare.

This is a challenge that many of the professional organizations that represent mental health counselors have been trying to overcome for many years. There have been changes, and significant ones, over the past five years. The problem remains, however, that there are veterans in need who are not receiving services because of their lack of ability to pay. Each of us, individually, has the ability to reach out to their local lawmakers and have a conversation about what changes need to be made to impact veteran mental health in our country.

Get involved in advocacy efforts. If you are a mental health professional, reach out to your professional representative organization and work with their advocacy branch. If you are a veteran, or support one, reach out to your local lawmaker or their veteran services liaison, and ask to have a discussion about what changes need to be made in order to reduce this particular barrier to mental health services. Change happens when enough of the population cares enough about a subject to be vocal about it, and this is certainly a change that absolutely needs to happen.

The (Inappropriate) Case for Suicide

When it comes to the veteran support community, there are fewer subjects, in my opinion, that are more sensitive than Veteran Suicide. There are also fewer subjects that are more individualized, and more complicated. Let me throw this firebomb of a question out there:

Are there some circumstances that make it acceptable for someone to take their own life?

Before you move on, take a moment and consider the thought that comes to your mind. Without judgment, without anyone knowing what thoughts you're having right now, consider your answer. It doesn't make you a bad person, if the answer is that there might be some circumstances where it would be acceptable. It means that you have a complicated opinion on a complicated subject.

Personally, just to give you my answers to the quiz, I don't believe that there are any circumstances in which suicide is acceptable. But I know that some of you do.

I've had this discussion before, when I lead Applied Suicide Intervention Skills Training workshops while in the Army[3]. In a portion of the course that helped participants understand their attitudes and values towards suicide, a similar question to the one above is asked, and a discussion about people's answers always proved instructive. For me, when people tended to admit that suicide might be acceptable under certain circumstances, I would ask about those circumstances. Some of that, I'll talk about in a minute. For those who said that it is never acceptable under any circumstances, I would ask them:

What about Ariel Castro? The world was shocked in 2013 when it was found that three women had been held captive and repeatedly raped by Castro for over a decade. He was convicted of kidnapping and rape and sentenced to over 1,000 years of correctional custody. For someone to have committed such prolonged torture and abuse, committed such detestable crimes, is the taking of their own life acceptable? Or understandable? Some would say, "Coward's way out." Some would say, "Poetic Justice." This type of situation in which someone takes their own life after committing a crime, this is one of those complicated factors. In the 24 hours before this article was written in 2017, the media widely reported on two separate circumstances of suicide. First, murder suspect, Steve Stephens, who was being sought by police after posting a video on Facebook of himself committing murder, took his own life while he was being chased by police. Second, convicted murderer and former New England Patriot Aaron Hernandez, who was found in his jail cell on the morning of 19 April, 2017, after he had apparently taken his own life. Was suicide acceptable in these cases? Your opinions are your own, of course, and I for one won't judge you for them.

What about those who are suffering with extreme pain or facing a known debilitating illness? This argument stretches back to my childhood, with Jack Kevorkian making his argument for physician-assisted suicide for terminally ill patients. My first roommate in the Army was a huge Chargers fan, and when Junior Seau took his own life in 2012, I immediately thought of him. Seau was found to be suffering from Chronic Traumatic Encephalopathy (CTE)[4]. CTE is a degenerative disease found in individuals who have endured a severe blow or repeated blows to the head. Common in football players…and possibly veterans. Another example, not that it's needed, is Robin Williams, whose widow believes that he killed himself as a result of a degenerative neurological disease called diffuse Lewy body dementia[5]. Some may argue that, faced with inevitable degenerative mental or physical illness, it might be a blessing for someone to take their own life, rather than endure the pain and suffering, and put their families through that.

The problem with each of these cases is that we can speculate all we want, but the answers…the "why"…ended with each of them. No definitive suicide note was found with any of them, no explanation. For some, no explanation would seem to be necessary. "Of course Stephens and Castro would take their own life, look what they did," some would say. Tragic as they were, the deaths of Seau and Williams might be understandable, knowing what they were facing. This is one of the ultimate tragic problems with suicides…the eternal unanswered question of "why" and "what could I have done?"

As a mental health counselor that works exclusively with veterans, I have to keep in mind that it is a distinct possibility that some veterans I work with are at increased risk of taking their own life. The causes are as numerous and varied as the circumstances that each veteran has experienced. I can only hope that I provide enough of a sense of trust and connection that they will choose to reach out for help before taking that final, irrevocable step. I also have to be aware of my own complicated attitude toward suicide in general, and veteran suicide in particular. I don't want to see a veteran homeless. I don't want to see them in jail. If only my wanting to be so made it so, I would be pleased…but I know that it will continue to happen. I certainly, clearly, and absolutely don't want to see a veteran die by their own hand. In as many ways as possible, as often as possible, I want to make that point.

I don't consider any circumstance acceptable for a veteran to take their own life. Even the most vile, heinous, criminal act, even the most degenerative disease. That's just my opinion, and I appreciate that others might have theirs. I know, however, that shaky breathless moment that exists after you've just found out that you've lost another brother or sister. Or family member. The world is forever irrevocably changed after someone you know has taken their own life.

What were the names of the veterans who took their own life yesterday? Or the ones who took their life on the day Robin Williams killed himself? Just because we don't know who they were, doesn't mean that they didn't impact the lives of those who knew them the way these high-profile instances impacted many of us. Taking the time to consider and understand your own complicated attitudes toward suicide can be beneficial in making a difference in those around you.

Welcome to the first Cross-Generational War

Never before in American history has there been a chance for a child not born at the beginning of a war that would potentially be able to fight in the same war as their parent. That day is rapidly approaching.

I've been having some conversations about the impact that combat deployments have on families. For the last fifty years, military service has very much been a family affair. My father served in Vietnam, as did two of his brothers and a brother-in-law. My younger brother enlisted; he joined the Army as I was leaving Baghdad in 2007, and was assigned to the unit that replaced us. Six or eight months after he joined, he was stationed in the exact same region I just left. On my second Afghanistan tour, his first, we were in country together. You hear of these stories; brothers and sisters serving together, fathers and children. It's especially true for Reserve and Guard units; our engineers in Afghanistan in '09-'10 were out of Indiana. The crew we went out with the most were literally cousins, and said that they were related in some way to almost a third of their unit.

Maybe it's always been this way, but we're on the verge of something new: a toddler on 9/11 could potentially join the military and serve in the same combat zone as their military parent. My daughter was born 18 months before 9/11, my Son was born just over a month before. Neither of them remember the event, but are intimately familiar with it's impact on my life, and our lives. With permission from my wife and I, my daughter could join the military today. My son could join as soon as he turns 17. Regardless of whether they will or won't, the potential is there. If they do enlist, or decide to commission or whatever, the fact is that they could serve in Kabul or Jalalabad or Baghdad, just as I did.

Never before has there been the potential for this to happen.

Any veteran I know has stories of families serving together. When I was in Rustamiyah, Iraq, a soldier from another company was requested to be at the helicopter landing pad at a specific time. No one knew what it was all about...until the scheduled helicopter came in, and the soldier's father, who happened to be the regional Command Sergeant Major, walked up and gave him a hug. I heard stories from my father's time in Vietnam, when his brother and brother-in-law overlapped some time; my uncle showed up on my father's base with a jeep that had been...acquired...and the two of them drove to Saigon to meet my other uncle when he came in country. I had a similar, but less potentially illegal, experience with my own brother. He was stationed at a camp south of Kabul, and I was able to get permission from my chain of command to catch a ride on the Ring Route, a set schedule of helicopters that would travel from one base to another, to pay him a visit.

We spent the morning together, having breakfast and catching up…before I got back on my chopper and my own base in Kabul.

The stories of families serving in the same theater of war are often more tragic than these. In his book, *War*, and the documentary, *Restrepo*, Sebastian Junger details one of the most heartbreaking stories of a family of service experiencing the ultimate sacrifice[6]. PFC Timothy Vimoto lost his life in the Korengal valley of Afghanistan, a member of 2/503 Infantry Regiment, 173rd Airborne. His father, Command Sergeant Major Isaia Vimoto, was the 503rd Infantry Regiment CSM on the same deployment. More about their story can be read here. In another example, The book *The Invisible Front: Love and Loss in an Era of Endless War* by Yochi Dreazen[7] chronicles the story of Major General Mark Graham and his wife Carol, whose two sons, Jeff and Kevin, lost their lives within nine months of each other. Kevin died by suicide as an ROTC cadet at the University of Kentucky, while Jeff was killed by a roadside bomb in Iraq. There is tragedy, loss and sacrifice in these cross-generational stories, but a difference is that CSM Vimoto and MG Graham were career Soldiers, and their sons old enough to remember 9/11. We are approaching a time in which the outcome is the same, but the circumstances are different.

The modern military is very much a family affair. A fellow veteran, Jeff Adamec, said that he had a conversation with an eighteen year old recently, and this young man said: "America is at war, but that's not really true. The military and their families are at war. America doesn't seem to know or care anymore because it's uncomfortable to know. And that's a tragic truth." The fact is, however, that there is an even greater truth in that statement: military families are at war. Husbands and wives. Fathers and daughters. Mothers and sons.

And soon, there will be children who were not born at the beginning of these current global conflicts who will be participating in them. What lessons have we learned? What lessons are there still to learn, the impact of sustained conflict on these two generations?

We are currently experiencing the passing of the WWII generation. The last living U.S. Veteran of WWI died in 2011 at the age of 110. The cross-generational impact of these conflicts will be felt for well over a century, into the 2100s. While this may mean many things to us as a nation, the communities we live in, and the narrative we bring, I believe we also have to be aware of the impact on individual mental health and wellness. We are entering into uncharted territory, and it would be best if we did so with caution and awareness.

You Break It, You Buy It: A Nation's Responsibility for Veteran Mental Health

It was recently announced that the Department of Veterans Affairs is going to start providing emergency mental health support for veterans regardless of discharge[8]. This is a huge step in the right direction for getting all veterans the support they need. This service is going to be provided will be time-limited, and require the veteran present in an emergent condition to their local VA Clinic, but something is better than nothing.

It's national news, and any amount of water is welcome to a thirsty man in the desert. What this does, however, is bring the topic out into the open to start a national conversation: what do we, as a nation, owe to those who served in the military? What responsibility does an individual citizen have to someone who volunteered to join the military? This can be a complicated discussion with many different facets. Here are some thoughts from the perspective of someone who is both a combat veteran and a mental health professional:

Many veterans reject the thought of someone "owing" them something

The fact that this is even a discussion, that the nation has a responsibility to "take care" of those who served, is often rejected by many service members. The military does not cause service members to become solo warriors, depending on no one but themselves, but there is certainly a sense of "in" or "out." Members of the team, who've been there, are able to be included in the support network; they have earned the right, because the service member has a mutual obligation to support them. My shield covers my brother, so to speak, just as my brother's shield covers me. This doesn't mean that they reject everyone else, but the fact is that there is an attitude of service that rejects the concept of being served. This is something that the veteran has to understand and come to terms with.

Many veterans reject the thought of being "broken"

"Broken" conjures up thoughts of being damaged, useless, worthless. "Broken" is what happened to that guy or gal in your squad who just couldn't hack it anymore, and went to sick call or dropped out of the run. No one wanted to be that person in the squad, the one who just couldn't be counted on...so the thought of "you broke it, you bought it" is likely to be rejected by the veteran. The truth is, however, that military service members have engaged in and endured some of the most mind and soul altering experiences, and repeated exposure to these experiences has an impact on both mind and soul. Those outstanding athletes at the Warrior Games? In spite of physical (and psychological) injuries, they thrive and persevere, and while they also would reject the concept of being "broken," they have sustained damage to their brain and body. Like it or not, reject it or not, it's the truth.

Some in the community will reject responsibility

I understand that there are many in society who will reject the thought that they have a personal responsibility to care for a veteran who has been impacted by their military service. "Well, they signed up for it, they knew what they were getting into," the argument goes. Or, "They have doctors and stuff that take care of that, right? What do I have to do with it?" Outside of this being a self-focused point of view, many who enjoy liberty and freedom to express those thoughts might not consider the price that had been paid for them to be able to do so. Sure, perhaps around the 4th of July, Memorial Day, and Veterans day, does the country start to say, "thanks" but it's a word that is often not backed by substance. Veterans served without any expectation beyond what they were promised for their service. Just because the expectation is not there, however, does not mean the obligation doesn't exist.

It's a complicated problem with no easy solution

The challenge in providing mental health support to all veterans, regardless of era, type, and manner of service, is complicated by many factors. There is a severe shortage of clinical mental health professionals in our country. Of those who are licensed and certified, a significant number of them are not familiar with the cultural aspects of military service[9]. Mental health practitioners, although having spent significant time and resources to be able to become licensed professionals, are often expected to provide services at little or no cost. Much insurance, including Medicare and Medicaid, do not support mental health services at the same level they do for physical health. All of this combined creates the challenge that those who are in the greatest need of support, such as those veterans with bad paper, are also those with the least amount of resources. Someone has to pay the bill for veteran mental health services, and it should not be the veterans themselves.

The decision between good paper and bad paper is often arbitrary

I've been in the military. I know what happens when a service member gets discharged for patterns of misconduct. The choice of whether or not a service member gets a discharge labeled "general under honorable conditions" or "general under other than honorable conditions" is quite often an arbitrary one that happens on their way out of the military.

Someone in the service member's chain of command, acting on their own sense of "right" or "wrong" has the ability to recommend to the service member's commander of the type of discharge to be received. While the military has taken steps to safeguard this decision, such as by instituting administrative review boards for those service members who have served for longer than six years, this does not apply retroactively to the thousands of service members previously separated under these conditions.

Community providers have been supporting "bad paper" vets for years

A final thought is that, as is often the case, organizations outside the government have stepped in to address an issue that impacts their communities. The United Way, for example, is a long-standing charitable organization that builds a coalition of charitable organizations in order to make measurable impacts in communities. They often work alongside and on the behalf of local governments to address challenges in housing, health, education, and income. Similarly, there are clinical mental health organizations in the community that have been providing support for veterans with bad paper for many years. National organizations, such as Give An Hour, The Soldiers Project, The Headstrong Project, and the Cohen Veterans Network are all currently providing clinical mental health counseling support to veterans regardless of discharge status. Similarly, many local organizations are providing this support, such as our agency in Colorado Springs and the Sturm Center for Military Psychology in Denver. My agency alone has been providing support for veterans with bad paper since 2012.

As with many other aspects of veteran mental health, this is a complicated issue. There is no single, easy, or comprehensive answer. Did some veterans discharged other than honorably deserve that designation? Absolutely, but they are by no means the majority. The same can be said about some veterans who received honorable discharges but conducted themselves in a dishonorable way. The purpose of this is to consider what we, as a nation, owe those who sacrificed much on behalf of our safety and security. If this is important to you, join the conversation. Do the research. Listen to the stories of those veterans who sacrificed, were wounded physically and psychologically, and consider the price that they paid. Then consider who is going to pay that price…them, or the nation that called upon them to do so.

Six Thoughts on Posttraumatic Stress Disorder

Post-traumatic Stress Disorder. Shell Shock. Battle Fatigue. Call it what you want, it's that combination of psychological and behavioral reactions to exposure to traumatic events that people see in combat veterans. There is a lot of discussion, whether at the water cooler, kitchen table, social media, wherever, about what it is. Who has it. What can it mean for those who do have it, those who live with them, and those who work around them.

There are a lot of misconceptions around PTSD, a lot of unknowns that cause people to jump to conclusions. Here are six quick thoughts that might help you understand what PTSD is, and how it impacts veterans.

Just Because a Veteran Has Been to Combat Doesn't Mean They Have PTSD

One of the stereotypes that many in society have is that exposure to combat, or even just combat deployments, must mean that a veteran has PTSD. Witness the high-profile events in which military service members committed egregious criminal acts: the sniper-style shooting event in Dallas, the multiple homicides in Baton Rouge a week after that, or the Airport Shooting in Florida. In each of these cases, the veteran's military experience, including deployment experience, was examined to determine if PTSD was the "cause" of their actions. Correlation is not causation, as any scientist will tell you, and the fact that a veteran deployed to a combat zone does not mean that they will automatically develop PTSD.

The fact is, there are a large number of individuals who deployed to a combat zone who never actually saw combat action. They were never shot at, they were not threatened with rocket or mortar attacks, their lives were never really in danger from enemy activity. That does not mean that nothing bad ever happened when not in active combat; certainly accidents happen. I recall, as a young Soldier in Bosnia, walking from our living area to the motor pool. Everyone walked the same path; we had arrived in the winter, so the snow was trampled and worn. Guess what happened when the snow melted? All of the unexploded ordnance started to appear on the ground. I don't want to make it sound dramatic and say that we had been walking through a minefield, but if the path had been ten feet to the right or left, someone would have gotten hurt or killed. The military is an inherently dangerous profession, so certainly mishaps and accidents can occur…but the individual who deployed to Kuwait in 2012 was not faced with the same danger as the group of men and women hanging off the side of a mountain in 2009. So, just because someone's a "combat vet" doesn't automatically mean they "have PTSD."

PTSD Is Both a Natural AND Negative Reaction

PTSD is as much a neurological condition as it is a psychological one. Studies have shown that repeated exposure to trauma results in physical areas of the brain, specifically the amygdala and the hippocampus, actually undergoing structural change[10]. Activation of the amygdala and hippocampus combined with reduced activation and volume in the frontal lobe of the brain are indicative of veterans who display symptoms of PTSD. Therefore, *the brain is adapting the way the brain should adapt to repeated traumatic exposure.* It is doing exactly what it should do when exposed to high levels of cortisol and adrenaline and constant activation of these regions. Many who rail against the "D" in PTSD say, "why pathologize a natural reaction to combat?" That's absolutely right...PTSD is precisely a natural reaction.

That doesn't mean that it's beneficial. There are hundreds of things that are natural reactions to our environment, and they're still not beneficial. Allergic much? Anaphylactic shock occurs after a bee sting, it's a reaction of one organism to a substance from another. Am I comparing PTSD to allergies? Of course not, but just because a reaction is typical for a small group of people doesn't make it beneficial. This very natural reaction can be a debilitating condition...it's both.

A Veteran Can Experience Challenges Even If They Don't Have PTSD

To be clinically diagnosed with PTSD means that the individual must meet a certain set of criteria[11]. But not all veterans meet all the diagnostic criteria at the same time. Some veterans don't experience nightmares or intrusive thoughts, but behaviorally isolate themselves. Or are extremely hypervigilant. At one point in their lives, or in one particular location, they can experience the mood symptoms but nothing else. Not only that, PTSD may not be the issue, but guilt or shame may be the issue. Or a lack of purpose and meaning in their lives, a substance abuse problem, relational difficulties, or any other of a number of challenges that many veterans...and many non-veterans...face.

A Veteran Can Develop PTSD from a Single Event or Multiple Events

Complex trauma as a result of exposure over an extended period of time can result in PTSD that is much different than that of a single event. Any one of us can experience PTSD after a vehicle accident or a natural disaster. As a matter of fact, these are two of the most common conditions that non-veterans would experience when it comes to developing PTSD. Complex trauma, however, occurs when multiple traumatic events happen over a period of time, or even a lifetime. As I've mentioned before, the military is almost a refuge for many who had a traumatic childhood, when is then compounded after exposure to trauma as an adult.

A Veteran can Develop PTSD Even If They've Never Been to Combat

I've worked with veterans who have developed PTSD, but have never deployed. The 82nd Airborne Division responded to Hurricane Katrina[12], and had to recover the remains of their countrymen. Just that level of devastation can have a traumatizing effect. I have known, of course, service members who have been raped. Any event that causes threat or actual death, dismemberment, or sexual violence is described as "traumatizing" when it comes to PTSD. Having a bad experience with a supervisor while you were in the military? Not PTSD. It could certainly lead to helplessness, frustration, angry outbursts, depression, and a whole host of other mental health concerns, but not PTSD.

But, as I mentioned above, the military is a dangerous profession. Parachute accidents, howitzer misfires, training accidents, each of these are potentially traumatic and could cause someone who has served to develop PTSD, even if they haven't deployed.

A Veteran can "Have" PTSD and Still be Functional and Productive

So perhaps a veteran has been exposed to multiple traumatic events and meets the clinical definition of PTSD. Does that mean that we should write them off entirely? Of course not. The condition can be managed to the point where the veteran may not feel the challenges for months, if not years. With treatment, the veteran can absolutely manage their PTSD symptoms to the point that they can be a functional and productive member of their community, their workplace, their family.

Understanding this condition is a key factor in reducing the gap in understanding between those who have served in the military, or love those who have, and those who have not served. What questions do you have about PTSD? What do you know about it, and how can we have a conversation that can help you understand more? Reach out, comment below, and join the conversation.

PART 2

Developing Personal Awareness of the Need for Veteran Mental Health

What is necessary to change a person is to change his awareness about himself –
Abraham Maslow

In addition to raising awareness in the community, a veteran will not reach out for support unless there is a measure of awareness within themselves about why they think the way they do. Why they do the things they do. Without personal awareness about those things that we carry over from our military service, we could possibly think that what we're going through is no big deal. That it's something that we can handle.

Along with developing awareness about why we think, feel, and act the way we do, we can and should come to the awareness that there are ways to change this if it's getting in our way. I often tell veterans I work with: if you want to go live in a cabin in the woods, and you're not hurting yourself, your family, or breaking the law, then go for it. Understand *why* you're doing it, though; if you're okay with your reasons, then go for it. If isolation is a desire for you, and isolation doesn't bother you, and your family and friends are okay with it, then more power to you. The challenge comes when we engage in these behaviors and we don't know why, or don't know how to change them if we don't like how they're impacting our lives.

This collection looks at how a veteran or military spouse can develop awareness about different aspects of their military connection, and how those aspects impact their post-military lives. Why veterans love their time in combat, and hate it at the same time. The way we think and feel has an impact on how we act, and vice versa. Gaining awareness about these things, and then deciding whether or not we like them, is the first step towards making a change.

Action and Avoidance, Then and Now

"The oldest and strongest emotion of mankind is fear, and the oldest and strongest kind of fear is fear of the unknown" H.P. Lovecraft

Veterans are among the most courageous people that I know. Dauntless. Unstoppable. The Army has established seven core values that all Soldiers should possess: Loyalty, Duty, Respect, Selfless Service, Honor, Integrity, and Personal Courage. For many years, I thought that Integrity was the most important, for how could I have any of the others if I didn't have integrity?

After 9/11, however, my thinking changed somewhat. Courage. For those firefighters to run up while everyone was running down…like Rick Rescorla[1], from *We Were Soldiers, Once…and Young*…that's courage. That's sacrifice. To run towards the sound of gunfire, instead of away. To stand firm in the face of danger, to even care less about your own death than you do about those you are with, and knowing they feel the same.

In the military, you have hundreds of daily chances to overcome fear. Hundreds of thousands of service members, like me, have decided that they would jump out of a perfectly good airplane. In all of my time in the military, I never saw someone refuse to jump, the cardinal sin of the Airborne community. Never saw someone freeze in the door, never a saw someone not do what they have committed to do.

Combat. When you take point on patrol, or lead vehicle on a convoy, or stack on the door as the breach man, you did it. When I was leading security escort, I had a team that was always lead truck…and, if I had someone else take the lead (not that I can recall such a thing happening), then they would want to know *why*.

What about calling for support when you were in the military? You didn't do this alone. Say all you want about the fobbits back at the command post, they were (mostly) there when you needed them. Intel? On point. Fire support? Whenever you needed it, if not always wherever. Snipers in overwatch? Yes, please. There was little hesitation to reach out to request help if you needed it, and you needed it often. One of the most amazing things to me was the amount of support I had in the air as a Platoon Sergeant in Afghanistan: Kiowas right above me, drones above them, F-16s above the drones, and, somewhere up on the edge of space, a B-52 bomber. Just to make sure me and my guys and gals were safe.

And then we come home.

How is it possible that I can jump out of airplanes, take point on a patrol, stand firm in the face of adversity, but yet start shaking in my boots at the thought of my first civilian job interview? How is it that I can get on the comms to call for support anywhere from a mile ahead of me to a mile above me, but I can't reach out for the assistance I need here at home? I'm still the same person, the dauntless service member. I can stack on the door of a compound in Sadr City, but I hesitate at the door of the therapist's office? Doesn't make much sense.

I've thought about this: I think it has something to do with confidence. In yourself, in those around you, in your support. And confidence is about certainty.

When you prepared to bust down a door in combat, you were confident. You didn't know what was on the other side of that door; at that moment, the immediate future is completely unknown. You know what *might* be on the other side of that door, or was even *probably* there. But you didn't *know*. And you went anyway. Why?

You were confident that you would be able to handle whatever it was. You had certainty, based on your training and experience, that what was unknown in the world was not greater than that which was known within yourself. You knew that your team had your back, and you knew it with certainty. The same goes for support; it was rare that you called for support if you were in contact and didn't receive it. When you called, you got what you needed.

When you stand in front of the therapist's door, however, or going into your first interview, somehow that certainty has gone away. You somehow have lost the certainty of your ability to handle whatever comes next. It doesn't matter that whatever is on the other side of the door here is nowhere near as dangerous, physically, as what was on the other side of the door there.

The same goes for your support. Somehow, we lose the confidence that someone will be there when we reach out for support. We hear the stories of the horrible service at the VA in Alabama, for example, and think that the same thing happens at the VA down the street. We hear stories in the media of someone getting an answering machine on the suicide prevention hotline, and say to ourselves, "well, that's another example of the fact that they don't care about veterans. Might as well not even try." We turn around and pack up our kit before we even roll outside the wire.

That's not who I am. That's not who we, veterans, are. The profession of arms is now, and has been for as long as I remember, a noble and honorable profession. Even if it seems as though some in our lives may not think so, it is.

When you were in the military, you had the courage to face death, because you were confident in your own ability to overcome it, the ability of your brothers and sisters to bear it with you, and the presence of support. Now that you're out, you can once again become confident in your ability to handle anything. There are brothers and sisters out there, veterans, who will be there if you need it. Support is there, all you have to do is ask and ask and don't stop asking until you get it.

Over there, you were courageous enough to face certain death. Over here, are you courageous enough to face certain life?

How Do You Hold On To Failure?

Do you carry your failures in your pockets, or do you have them tattooed on your skin? Which this is for you can be the difference between peace or chaos in your life. Some of us wear our failures like a second skin, the scales of a dragon that become a sort of comfortable cage. Others of us recognize that mistakes and failures are ours, and we own them, but they are not a part of us. They don't make us who we are. Sometimes, the belief that we are our failures is buried so deep that the core belief has become the piece of sand that lies at the core of a pearl. Our entire identity is built on the idea that we're a screw-up, a bum, that we deserve what we get.

When you get to the place where your failures become your identity, you're so far down a hole that you don't even realize it.

It doesn't have to be that way. Here are four things for us to remember in order to separate ourselves from our mistakes and failures, so that we may grow. Learn. Be a better version of our current selves.

We Are Not Our Mistakes or Failures

Making a mistake is something that humans do. Failing at something is what makes us learn. That's not to say that we must fail to learn…we can learn from the failures of others, after all. But if we make our failures a part of us, it can keep us from learning, keep us from success. We can become trapped in a cage of our own construction, the mistakes and failures of our past holding us down like ropes and chains. Who we are is a result of the entirety of our lives, from birth up until this current moment, but we are not any more of a success because we once knocked a ball out of the park as we are a failure because we couldn't hit it past the pitcher's mound. If we tend to make our failures our identity, then we will have a tendency to look at ourselves in the mirror and see our failures written on our face in invisible ink. We stare into the eyes of the guy or gal looking back at us, and assume there's something wrong with them. We carry blame and shame when blame and shame do nothing for us, but make us feel worse.

Remembering Our Mistakes is Not Reliving Our Mistakes

We've all done something cringe worthy. We all have those embarrassing moments that we wish people would forget, but they don't. Worse than that, we have monumental failures, huge, life-changing mistakes. We did something to someone. We didn't stop something from happening that we think we should have. Or we could have. We screwed up, royally, and we paid the price. Perhaps someone else paid the price, which is worse. It's important to remember these mistakes, because if we forget them, or we're not aware of them, we may make them again. We can't help others learn from them. There is a big difference, though, between remembering the mistakes, acknowledging them, and reliving the mistake over and over again. That's the difference between keeping the mistake in your pocket or tattooing it on your skin. If it's in your pocket, you can take it out. Look at it. Consider it, what lead up to it, what the consequences were. You can put it on the table, separate from you, and observe it from different angles. We can detach the mistake from our sense of self, and remove the emotion from it. Then we can truly learn from it.

Not Reliving Our Mistakes is Not Absolving Us of Responsibility

Sometimes, we feel as though our mistakes are so horrible that we have to relive them, over and over, like a terrible unending nightmare. If we don't keep doing that to ourselves, then it must mean that we're not accepting responsibility for them. Since when does accepting responsibility for a mistake or a failure require us to relive them? Can't we accept them, forgive ourselves, move on, AND acknowledge our failures? That we are flawed, fallible...human? Sure we screwed up. Probably big time. AND we can learn. Continuously beating ourselves up for a mistake in the past is counterproductive to healing. Would you re-break a broken arm, just to remind yourself not to break it again? No. You mend. You heal. Maybe not perfectly, probably with scars, but you heal.

Giving Power to Labels Takes Power from Ourselves

Loser. Trash. Garbage. Monster. Pathetic. The things we call ourselves are worse than we would ever let someone call someone we loved. If someone called my daughter, or my wife, the things I call myself, I'd punch them in the face. Why, then, do we say these things to ourselves? We are people who have lost, we're not losers. We may have done horrible things, but that doesn't make us horrible. We may have chosen to act like garbage, but that doesn't mean that we are garbage. It's a fine line, of course, but it's the difference between continued pain and suffering and eventual healing. It's a small shift in the way that we think, but the shift of one degree at the beginning of a journey can change our eventual destination by miles.

It's a deliberate focus on keeping our mistakes and failures separate from who you are, from your core sense of self. It's a focus on making sure that you learn from your failures, that you grow, instead of continue to beat yourself up.

You make mistakes and have failures, you are not your mistakes and failures.

Is Anxiety about Shadows and Legends Getting In Your Way?

The Legend of Billy The Kid killed more people than Billy ever did

When we were preparing to deploy to Afghanistan in 2009, our unit was going through the typical spin-up exercises. Field exercises, where we went to remote training areas on our base for days and weeks at a time. Everything culminated in a month-long visit to the Joint Readiness Training Area located at lovely Fort Polk, Louisiana. As the Platoon Sergeant, I was typically the last one asleep, and the first one awake. In a moment of brief relaxation, one of my NCOs said, "You know they think you're a robot, don't you?" Apparently, my habit of going as long as I could fueled by coffee and edgy irritation had some unintended side effects. Maybe it was good to think that the leaders taking you into combat are somehow infallible and indestructible, but back then I was as human as anyone else, and remain so today. Sometimes, the legend grows bigger than the individual.

I've experienced this with my own leaders. My first squad leader has developed into some mythic hero, striding across the Army with an M16 in one hand and a steering wheel in the other, some incredible mix of Rambo, Buddha, and Patton. Depending on what story I tell, he either shrinks, as he's grabbing me by my collar and yanking a knot in my butt, or he grows, towering over me as the Lord of Discipline that I remember him to be.

I hope I never meet the real guy, because I'm sure to be disappointed.

Anxiety And Events

Sometimes, impending events can seem more daunting than they actually are. The shadow of an event, the unknown nature of the outcome, is enough to bring shivers to the knees of Hercules. Figuring out how to continue, to push on, despite those concerns, is what will bring success rather than failure or stagnation.

Anxiety is beneficial, to a certain extent. It helps us, it keeps us sharp, keeps us on our toes. I recall the advice of my Platoon Sergeant when I was in the 82nd Airborne Division. "You damn well better be scared jumping out of an airplane," he used to say, "It's when you stop being scared that you start being stupid." Having an appropriate amount of anxiety helps you to know that what you're about to do is important, so pay real close attention to it.

It's when anxiety turns into real fear, or even terror, that it becomes harmful. When anxiety causes you to freeze instead of react, or to quit before you even begin, it's time to take a real hard look at what you're afraid of. Is it the shadow of the event, or the actual event? Is it the legend, the tale, that you're reacting to, rather than the actual thing? Recognizing that you're jumping at shadows instead of what's really there can help calm you down.

Anxiety And Transition

About a year before I retired, I was talking to a friend who was the 1SG in one of the other companies in our Battalion. As we walked into her headquarters, she looked over at one of her Soldiers, a Staff Sergeant (E-6) and told him, "Get out of here! Go take care of what you need to take care of!" As we went into her office, I asked her what the deal was; she told me that he was going on terminal leave for his retirement in less than a week. She had been doing her best to make him leave, to get ready to transition, but he was still the first one in the office and the last one to leave. "The Army's all he knows," she said, shaking her head, "He's going to have a hard time when he gets out." For that Soldier, the shadow of life without the Army was huge and daunting, and he wanted to cling to what he knew as long as he possibly could.

How many times has that happened for you? The first actual interview you walked into, the first job fair. The first day on the new job. The night before you took over a new position. The day before a big presentation, a critical meeting. Even the big life moments, your wedding day, the birth of your child. These are huge life-changing events, or at least have the potential to be so, and so it's probably good that you're anxious. That you're on your toes, making sure things are going right. It's what keeps you sharp.

If you don't experience anxiety, then good for you. Maybe you're the robot that my joes were looking for. But if you're like the rest of us, the living and breathing people. There is always room for a healthy dose of anxiety, and never room for an unhealthy dose of it. It's not the lack of anxiety that is important, but how you react to it. Do you let it control you, or do you control it? It's by harnessing our anxiety, and using it, that we can truly get things done.

The Black Dog of the Veteran Emotion

I've often heard the psychological aftermath of military service described as a black dog. The black dog is an emotion, or series of emotions, that the veteran experiences, sometimes under control, sometimes uncontrollably.

The black dog of PTSD can be terrifying. It's always on alert, hackles raised, mouth curled in a snarl, as a warning. Warning others not to approach, as danger is near. Warning others to stay away, as they may BE the danger. Snapping, barking, growling, scratching at the ground, prowling back and forth but never taking its eyes off of you. And everybody. And everything. The black dog of PTSD has seen things and had to do things that no dog should ever have to see or do, and is constantly on alert as if it were to happen all over again.

The black dog of anxiety is pitiful. It's a dog that has learned that every movement contains danger, and even the slightest gesture can make it flinch. Tail tucked, cowering down, a mere shadow of what it once was. The anxiety dog has been kicked by a cruel master way too often, and looks out at the world with fearful eyes. Like a dog reacting to a thunderstorm, the black dog of a veteran's anxiety will react to other things: crowds, traffic, uncontrollable situations. The reaction will be similar as well. As a dog is afraid of thunder, and starts to tremble uncontrollably, pacing the room, so too does the anxiety dog. As our dog does, so do we.

The black dog of depression sits on the veteran, heavy and immovable. It weighs you down, holds you back, keeps you stuck and stagnant. The depression dog is sometimes a constant companion, always there, always present. It is as common as a shadow, enveloping the veteran, cloaking them. It curls at the veteran's feet, and in the veteran's heart, in the morning, follows their footsteps all day, and keeps them companion through the night.

The thing about the black dog of the veteran's emotion: it's our dog, and we love it. We don't always enjoy it, we sometimes wish it wasn't there, but it's comfortable and familiar. We might accept it as something that will never go away, or we might take it with us wherever we go as a matter of pride. *"Of course I have a black dog,"* we say when others point it out, *"I'm a combat vet, what do you expect?"* We even feed the dog, nurture it, allow it to grow more anxious, more depressed, more angry.

The other thing about the black dog: it doesn't have to stay that way. We can learn to control it, to make it less present and impactful on our daily lives. We don't have to make it go away entirely, as we are never the same once the black dog comes into our lives. Instead, we can train the black dog to be something else. The black dog of PTSD can be tamed. The anxiety dog can be calmed. The black dog of depression can be uplifted.

We can learn to understand the warning signs of when the black dog starts to come around, and manage how we react to it. We can understand the things in our environment that will bring out the PTSD dog, or trigger the anxiety dog. We can protect against those thoughts that allow the depression dog to show up.

We can also understand that, sometimes, the dog's gonna get loose and go tearing around the neighborhood. That's okay. It doesn't have to be often, and it doesn't have to be for long. We can learn to control it, manage it, understand it, and be okay with it. We can either resist the dog, be angry at ourselves because we are dragging the dog around with us, and even grow to hate the black dog of our emotion as much as we are comfortable with it.

We can do all sorts of unhelpful things to try to get rid of the black dog; we drink. We fight. We give the black dog of our emotion control over our actions, and hurt others. We may even hurt ourselves; because letting the black dog be in control of our lives is easier than trying to control it.

It doesn't have to be that way. We can come to a place of acceptance, of understanding, that the black dog of our emotion is part of us. It's who we are; it's *what* we are now. It's a matter of becoming aware of the fact that the dog is there, being aware of the impact that it's having on our life and those around us, and beginning to understand why. Why the dog is there in the first place. Why it sticks around, despite our best efforts to make it go away.

Once we become aware, and then understand, then we can choose to change. To tame the black dog. To get it under control. We don't have to do it alone; there are professionals who can help us calm the black dogs inside of us, just as there are professional trainers who can help train our real pups. You just have to find one who understands the black dog, and the dog's owner; they're out there, and while searching may be challenging, it's ultimately worth it.

Learn to control the black dog, don't let the black dog control you.

The Combat Addiction Paradox

"Compared to war, all other forms of human endeavor shrink to insignificance. God help me, I do love it so." – George S. Patton

I've previously written about how I hate war as only a warrior can[2]. Something I wrote in there, however, was overshadowed by the point I was trying to make. The part that was overshadowed: while I hate war, I loved combat. Like the paradox of the veteran story and the violence of action paradox, this is another unresolved imbalance. Many veterans feel this way: they'd go back if they could.

This is a topic that has come up recently in several discussions I've had. How can we hate war, hate killing, hate the psychological impact of it…but at the same time love it, long for it, crave it? I was having a conversation with a colleague, and they asked me a question that I hadn't considered before.

"If you could close your eyes, and in an instant, open them and be anywhere in the world for 24 hours, where would you be?"

I thought about it, briefly. I'd been to London. North Africa. I've been on beaches and mountaintops. As I was rolling each of these foreign and exotic places around in my mind, I realized what the truth was: I'd choose to be in the Kunar River Valley, Afghanistan. In a heartbeat, without hesitation.

It was there that the most significant times of my life were spent. Outside of the moments that surround my family, such as my wedding and the births of my children, and the spiritual moments relating to my Christian faith, the most important moments of my life occurred on the roads winding through those mountains.

As Uncle George said in the quote at the beginning of this article, sometimes things in life after the military pale in comparison to our experiences in combat. The rush, the adrenaline, the camaraderie, the skillful execution of abilities developed for just this moment.

While not the cause of PTSD, this is a common attitude held by many combat veterans.

Veterans talk about being addicted to combat. Being a mental health professional, I feel as though it's my duty to specify that the only classified clinical mental health disorders that qualify as addictions focus on substances…alcohol, amphetamines, opiates, etc…with only the exception of gambling disorder. Other "behavioral addictions," such as compulsive sexual behavior or compulsive exercising, are not considered "addictions" in the clinical sense[3].

But oh, yes, how we love it so.

The challenge with any obsession, or compulsion, or whatever you want to call it, is that it overshadows our current lives, and we tend to long for the glory days of years gone by. We forget how hard those glory days really were, and instead pine for the bygone days when we were younger, stronger, slimmer and yes, had more hair.

We become stuck in the fact that who we are is a lesser version of who we were, and we tend to walk through life backwards. Is it a form of the high school football star who never really made it big in college, and he goes back to the high school kegger in his too-small letterman jacket?

Don't get me wrong. It's great to reminisce. To reflect on where we've come from, in order to understand more about where we are and give direction to where we're going. But to long for the good old days? To chase the rush and play the same song in order to capture lightning in the bottle just one more time? We are farther along on our life path than we were then. We cannot step into the same river twice.

I tell veterans all the time: we are now and forever will be veterans. I can't go back and be the "old me," the 17 year old kid that knew nothing and had less going for him. I can't even go back and be the 25 year old me, the crazy fool full of piss and vinegar, jumping out of every aircraft he could get onto. Similarly, we can't be something we're not; although I no longer wear a uniform, I'm not a civilian. Because I no longer wear a uniform, I'm no longer a soldier. I'm a strange mix of the two, a thing called "veteran," which is hard for both service members and civilians to relate to. And even more, a *combat* veteran, one which others who have served in the military but who have not been to combat sometimes find it hard to understand.

So a veteran wants to retreat into their past, but can't. So some do the next best thing: retreat into their memory of the past, which is a hazy and imperfect thing. They long for the days, the horrible glorious days, where bullets flew and the smell of copper and carbon filled the air. They crave those days as much as anyone craves anything; for some, no lover, no substance, no situation can ever measure up to the dizzying heights of combat.

But it doesn't have to be that way. Our lives did not peak on a mountaintop in Afghanistan, or a jungle in Vietnam. They do if we think they do; if we see that as the ultimate achievement in our lives, and everything else is downhill and meaningless, then that's what we will believe. That's not the case.

If you find yourself stuck here, then reach out. Find a counselor or a therapist that understands what you're going through, and sit down and talk.

You might find that the best days of your life are truly in front of you, not behind.

The Dangers of an Offensive Defense

"Offensive operations, often times, is the surest, if not the only (in some cases) means of defense" – George Washington.

In other words, the best defense is a good offense. In my experience, however, veterans often deploy a kind of offensive defense in order to not let anyone get close, when that may often be the best thing for them.

There are a lot of reasons why a veteran thinks it may be better to be bristling with spikes and shards than to let someone close. They've been hurt too often, emotionally or even physically. They don't trust others to understand what they're going through, and don't believe that they can truly be helped. They'd rather be alone behind stone walls with a bristling defense than have to understand what they're going through.

The Effectiveness of the Bristling Defense

The thing about the claws and knives defense is that it works pretty well. It usually provokes one of two reactions: the shocked withdrawal, or the vigorous counterattack. I experienced this recently; I was having a conversation with another professional, who is also a veteran, who started off the conversation with hostility and judgment. About nothing in particular! I was somewhat surprised at the level of anger that seemed to come out of nowhere, and started to backpedal for a minute. This is one of the reasons that the aggressiveness is effective: the veteran has their "opponent" off balance, and eventually the other person says, "forget it, I don't have to put up with this." This feeds into the veteran's idea of, "nobody gives a crap about me, it's better for me to be alone anyway."

A vigorous counterattack, which also comes as a result of the bristling defense, plays into the hands of the veteran as well. "AHA!" the veteran says, "they ARE against me!" Having this type of coping mechanism is a self-fulfilling prophecy, in which the veteran sets themselves up as a target for the other person's frustration, and then holds the other person responsible for engaging the target. In my opinion, if you don't want someone to press a particular button, you don't put the button in the middle of the console and illuminate it with a flashing sign that says, "don't press this button." You're sort of asking for trouble.

A Double Edged Sword

The problem with the bristling defense is that it does harm to both parties. If we see nothing but enemies around us, we're going to treat them like enemies. If they weren't enemies before, they certainly are now, and who needs more enemies? By keeping everyone at arm's length (or spear's length), the veteran may think that they're keeping themselves safe, but they're paradoxically harming themselves. It's through relationships that we communicate, and especially through relationships that we engage with society. If we consistently drive everyone away, then we will be systematically isolating ourselves, without an opportunity to express and examine our thoughts to see if they're valid.

The bristling defense builds a reputation that perpetuates isolation. "Man, don't deal with that guy. He's a jerk." The word gets out that he's difficult to work with, or abrasive to others, or unpleasant to be around. With so many cool people in the world, who wants to hang out with a jerk? Again, this plays into what the veteran truly wants, which is to be left alone, rather than what the veteran truly needs, which is engagement. We can only hang out with someone that throws elbows for so long, without saying "forget it" and walking away.

Awareness is key

Like with just about anything else about ourselves, it starts with personal awareness. I may not be fully aware of how abrasive I can be, or how I've put up a bristling defense in order to keep people out. The claws come out automatically, rather than with conscious effort. Again, the problem is that nobody sticks around to try to help the veteran become aware. It usually takes a crisis moment, or someone sticking around even after getting blasted, to help the veteran come to a place of awareness.

The problem then becomes, if someone does stick around in spite of the bristling defense, that the veteran doesn't know how to handle it. Again, in my interaction with my aggressive veteran colleague, the claws came out and then slowly went back in. I didn't retreat, and I didn't counterattack, I simply patiently and repeatedly explained myself. "Well, you seem to know what you're talking about, but they sure as hell don't," they said. The bristling defense seems to get the retreat or counterattack response so often, the veteran is not certain what to do when someone does neither.

If you find yourself isolated, it may be worth it to take a look at what you're doing to create that isolation. What are you doing to make it happen? Is it helpful, does it work for you? Does it make you feel good, or is it pleasant? Likely not. There's probably a lot of anger, and mistrust. The key is putting the claws away and letting someone in the perimeter. Find someone you can trust, who isn't going to let go when the spikes come out again. Once you learn to trust, then you can start to heal.

PART 3

Developing Resilience to Recover

"Out of massive suffering emerged the strongest souls; the most massive character are seared with scars." Khalil Gibran

If we never face adversity, then we will never know how to react to it when it happens. That's the entire premise of basic military training, in whatever form it takes: putting people in stressful and challenging situations that they haven't experienced before. It is done in an incremental way, from small challenges to large challenges, allowing individuals and teams to build, then bend, then build again. This is something that veterans are intimately familiar with from their time in the service. *If it ain't rainin', we ain't trainin'* is a phrase that has probably been used by noncommissioned officers since the beginning of time. George S. Patton said, "a pint of sweat prevents a gallon of blood," meaning the effort that we make before the battle can positively influence the outcome.

The alternative to resilience is capitulation. Breaking when we fall or bouncing when we fall can make a significant difference when it comes to success in our post-military lives. Tapping into the resilience and strength that we had when we were in the service can help us recover more quickly and to higher levels than we were before the thing that took us out. Resilience is a key factor in posttraumatic growth; after extremely stressful and traumatic events, we can become better and stronger versions of ourselves.

The following collection talks about resilience, and bouncing back, and tapping into something that helps us respond to adverse life conditions. The thing about resilience is that it can be developed just like any other skill. We first have to recognize the need, then develop the ability, and the more we apply it, the stronger it becomes.

Developing a Resilient Response to Attack

There are some psychological concepts that everyone just kind of "knows." Things like Freud's Oedipus Complex (which is offset by the less well known Electra Complex)[1] or Maslow's Hierarchy of Needs[2], which is discussed more in business than in psychology. Another well-known concept comes from the behaviorists: an individual's automatic response to an external threat: fight or flight[3].

These reactions are all around us. Is someone looking at you wrong? Let's gear up for battle. Or maybe let it get to you, and drag you down. An angry "What's wrong with you" is a psychological fight response, just as "What's wrong with me" is a psychological flee response. These thoughts can impact our moods, which in turn impact our behavior.

Consider walking through a crowd, and everyone has a bat in their hands. You have one too, but you can control your bat because you have the awareness of the damage it can cause. Some people have a little t-ball bat, and others have a great big ash wood Louisville Slugger. Some people are swinging their bat around without really understanding that they're even doing it…poking, jabbing, and thrusting indiscriminately. Other people, unfortunately, are deliberately targeting others with their bat. You seem to have two choices: get the heck out of there, retreat into isolation, or start laying into everyone else around you with your own bat. Neither is particularly helpful. Here are some thoughts on how to change how you react to an attack.

Whichever Reaction You Have Gives Power To The Threat

Our reaction gives validity to the attacker. If they don't realize what they're doing, okay, maybe we don't bash them too hard. But if they're attacking us deliberately? We get to retaliate against them, right? My response: it's their problem; I don't need to make it mine. If I buy into what they're doing and retaliate, I'm making it my problem. Sometimes, when someone criticizes us, we accept it without any further explanation; it sticks to us like Velcro. Other times, when someone criticizes us, we react with anger. Either way, by attack or retreat, you're giving power to the one doing the attacking.

You Can Choose Your Reaction If You Give Yourself Time

Consider, instead of attacking or retreating, standing your ground. Making yourself impervious to other attacks. Consider an old oak tree. If someone walked up to it and started bashing it with a bat, who's going to get tired first? Sure, the bark may sustain a couple of nicks, but the core of the tree is not impacted. Or maybe a ten foot tall stone wall. A bat against stone…might as well be a storm against a mountain. The storm, the attack, will eventually run out of energy, and even those who are being deliberate with their attacks will move on to someone who responds, because that's often what they're looking for. By standing your ground, not attacking and not retreating, you are being rooted firmly with stability. You can then help others around you, those that you care for, to do the same.

Establishing Defenses That Can Go Up Or Down

So maybe those in the crowd who are swinging their bat are those that are closest to you. You don't want to retreat from them, and the only reason we want to attack is to protect ourselves. What do we do then? Make the defenses mobile. Storm coming? Shields up. Storm subsided? Shields down. Let's talk about what happens when the storm hits, for both of us. Let me caveat here…if anyone is in physical danger, then remove yourself immediately. I'm talking about attack and defense in the metaphorical sense, not the literal sense. I'm referring more to emotional defense than physical defense; no one, ever, should endure abuse, intentional or otherwise.

Fight or flight. Natural responses. Not our only choice, though. We can stand our ground and protect our core self. By building resilience, we can better weather the storms of life.

The Moment After the Punch

There is always a moment that exists after getting punched in the face. What we do in that moment, and the moments beyond that, defines us. For better or worse.

So maybe we're not being punched in the face on a regular basis, but there are certainly left hooks and body blows that life, and other people, send our way. Relationships ending. Jobs ending. Plans being subverted. Betrayal. Insults. Rejection, in many different ways, of our ideas, our dreams, our thoughts, sometimes even of who we are.

It's that breathless moment, where the wind is knocked out of you and your ears are still ringing, that seems to stretch out into eternity.

But the second hand doesn't stop. Time continues to move, and we can move with it. We are on top of the world one minute, and the next minute, we're flat on our back, looking up at the sky, wondering what the hell just happened. Maybe we feel shock, or betrayal, or anger, or fear. The impulse may be to jump back up and counterpunch, to respond to an attack with an attack. The impulse may be to remain lying on the ground, too defeated to get up. It is in that critical moment that we have the power to influence our future.

If you haven't read my work before, I am a huge fan of Viktor Frankl[4], one of the eminent neurologists and psychiatrists of the 20th Century. This was a guy who knew what it was like to take a shot in the face by life; he was a concentration camp survivor. He lost his first wife in the camps, saw countless of his friends perish, and witnessed the ultimate evil that humans can do to each other. Frankl said this, among many other things:

Between stimulus and response there is a space. In that space is our power to choose our response. In our response lies our growth and our freedom.

Between that punch in the gut, that betrayal, that disappointment, and our reaction, there is a space. That space can be a split second, or it can be a year. Widening that gap between stimulus and response gives us the ability to choose growth, or to continue the chaos.

I've failed in that space. I know that failure is a judgment, but in this context, failure in my mind was actions that I took that made the situation worse, not better. Remember the story about that baseball on my desk[5]? I've thrown it before. Out of anger. I'm not talking about an underhand lob or a leisurely sidearm throw to my son in the backyard. No, I'm talking about a full-on, Nolan Ryan windup and release. The control wasn't there, but the velocity certainly was, and it left a dent in the door that it hit. Not saying this to brag, because it's certainly not something I'm proud of, but to illustrate the point that I had a choice in that moment between the event that made me angry, and my response to it. In hindsight, I chose poorly.

And there was a minute that existed after that explosive moment. And a minute after that.

We have the ability to make choices that can make the situation worse, or we can choose to salvage the situation in some way. To make it better. To be better, to be stronger, to be better versions of the same old us.

If you get knocked off the hill, and it's going to happen, you know that you've been to the top of the hill before. If you did it once, you can do it again.

But it's not going to happen if you keep lying on your back.

The Path of Painful Truth

Truth hurts. Maybe that's why we don't use it much. I don't mean that everyone is walking around with lies in their mouths, but how often do we show our real selves? It's dangerous. And painful. And a bad idea. Others could use your truth against you, and in a world where the edge is everything, we can't afford to show our real selves…can we?

There is something therapeutic, unburdening about walking the path of painful truth. After some time, it makes us feel like a weight has been lifted. It makes us seem more alive. We're no longer carrying secrets. We're no longer hiding something from the world. I've been amazed, at times, at how easy it is for a veteran to come in and talk to me as a mental health counselor. In many ways, even most ways, I'm a complete stranger. They've never met me before, I'm no one to them. Sure, maybe they've heard about me from another veteran, or read something or watched something or listened to a podcast. But at the end of the day, they don't *know* me.

One of the challenges is the fact that we are so used to not-truth. I don't mean fake or false, but fantasy. We're used to watching movies and TV shows where no one ever says "um" or "uh," or comparing ourselves to someone else's online persona. That's when we start to doubt ourselves, our ability to accomplish what we want. We have to present ourselves as this *perfect thing*, which doesn't actually exist.

By way of example, my buddy and fellow Change Your POV Podcast Host, Eddie Lazzari, engaged in a bit of honesty on one of his shows[6]. He shared a conversation that he and I had, offline, about his need to watch war movies and shows as a kind of atonement for what he did and what happened while he was deployed. It was a very public, and personal, revelation of a personal truth that he wasn't happy about.

Do you have to do honest the way Eddie did, publicly? Of course not. I didn't realize he was going to talk about our conversation, as much as I didn't realize how much it impacted him. I also have my own version of a painful truth that I've shared. One of the many reasons I became a mental health professional? I saw how much service in Vietnam impacted my father. I saw how it impacted my brothers and sisters, and I saw how much it impacted me. Want honest? Check out what I wrote about me being a Wounded Healer.

So how do you do it? Here are some steps you can take to get on the path:

Start By Being Honest With Yourself

Honesty isn't going to get out if it's not in you first. A lot of times, this starts with awareness; as Eddie mentioned in his podcast, he had nothing against veteran mental health, but never considered it for himself. Once he got talking, though, it was hard to stop. Being honest about what bugs you, what doesn't bug you, all starts with self-reflection. It's often been said, the worst kind of lies that we tell are the ones we tell ourselves. We tell ourselves we're ready for something, when we're really not. Alternatively, we think we're not ready for something, but we actually are. We are oblivious to how our spouse, or our children, or our parents are feeling. The first step in harnessing the power of painful truth? Realize that there's truth in you to be told.

Share Your Truth with Someone

I consider every moment given to me by a veteran as an honor. Trust is a key component of what I do. Eddie had a thought, like he said, and called me up out of the blue one day. He might not have thought of talking about some of this stuff when he called me, but at one point, he decided to go ahead and speak some stuff out. Because of trust. Our shared deployment experiences, which included combat and shared location, might have helped, but somehow there was enough trust there for him to bring it out.

That's the next step after self-reflection…sharing with someone you trust. Maybe it's your spouse, and, then again, maybe it's not. Sometimes, it's easy to be honest with someone you *don't* have an intimate relationship with. Several years ago, I participated in a group where military spouses could come ask questions about their service member's mental health. They could ask me questions that they thought they couldn't ask their husbands…and I could answer them, because they weren't my wife! It might have been a hard conversation to have with someone I love, but easy to have in a non-judgmental environment. Regardless of who you talk to, getting it out is the important thing.

Push Through the Pain

Just like getting back into the gym, or hitting the track again, walking the path of painful truth is going to bring on a bit of soreness, but you know that it's part of the price to reach your goals. Sometimes the truth is painful to you, or to others. Sometimes just *acknowledging* the truth can be painful, because we're not used to it. It's like using a different set of muscles. A while ago, while I was still active duty (and about ten years younger, and much, much lighter) I used to run one half-marathon a year. Once, I was on temporary duty in Delaware, and I was training for a half marathon in Annapolis; I was probably running thirty miles a week or so. The unit I was with played a pickup game of football one day, and the next day I woke up *sore!*

I used a totally different set of muscles in an entirely different way…the stop-start and quick takeoff of football was much different than what I was used to. I wasn't used to it…but I pushed through it. Just like Eddie did in his podcast, it was uncomfortable to talk about this stuff…but he didn't let the discomfort stop him.

Get to a Place of Peace

This is where you will eventually get. Peace. A post-military life not filled with pain or discomfort, but one of peace. If you're no longer in the military, your war is over. The current wars, and those of the future, belong to someone else. Spending time hoping for another shot is an exercise in futility, and a sure way to disrupt peace. You'll find that, once you start being honest with yourself, sharing with others, and pushing through the pain, you will get to a place where the ghosts no longer haunt you. That's the ultimate goal. You've served your watch, and done so with honor. No need to continue the struggle.

Veteran Mental Health and the Balancing Power of "If-"

When talking about veteran mental health, I often talk about the need for balancing different aspects of their lives. I've written about finding balance before, specifically looking at how to find equilibrium in your physical, mental, emotional, and spiritual wellness.

Another aspect of the need for balance is to be able to have equal measures of head and heart, to balance between thought and emotion. Up and down, rational and emotional, each of these are needed to be able to communicate effectively, connect with others, and interact with the world around us. We can't all be Mr. Spock or Sherlock, any more than we can be all Captain Kirk or John Watson. It takes both to make us whole people.

This also applies to our relationships. Spock balanced out Kirk and Watson balanced out Holmes. Like them, we often bring balance to others in our lives. This is especially true when it comes to chaos, strife, or conflict.

We can overcome conflict by acting in an opposite manner to those around us.

Have you ever noticed that you can calm down someone speaking loudly by speaking softly? Or counterbalance someone who is speaking quickly by speaking slowly? By taking the opposite approach when others around you are creating chaos, you can bring peace and order to your relationships.

This isn't a new concept; as a matter of fact, one of my favorite authors wrote about this exact thing in a poem to his son.

Rudyard Kipling, if you're not familiar with him, was a British journalist and author during the late 19th and early 20th Century. He is best known for The Jungle Book, but if you're a veteran of the Afghanistan conflict and you haven't read The Man Who Would Be King, you might want to put that on the to-do list. It's a story about western Soldiers and their experiences in what we know as RC East, Nuristan, and the Kunar-Korengal region of Afghanistan. While there may be a discussion about this in the future, today I'm talking about Kipling's poem[7] "If-" While not a veteran himself, he is what is now known as a Gold Star father; he lost his son, John Kipling, to whom he wrote this poem, at the Battle of Loos in WWI. Here are some thoughts on how the opposite action observed in If-can be applied to veteran mental health.

Veteran Mental Health Means Keeping It Cool When Others Don't

IF you can keep your head when all about you

Are losing theirs and blaming it on you,

If you can trust yourself when all men doubt you,

But make allowance for their doubting too;

If you can wait and not be tired by waiting,

Or being lied about, don't deal in lies,

Or being hated, don't give way to hating,

And yet don't look too good, nor talk too wise:

This is an area that I often fail at. I describe it as allowing my passion to overcome my effectiveness. We saw much of it in combat, or in the military in general...when people get heated, things get missed. Mistakes get made. If you can remain calm when others around you aren't, or can trust yourself and what you've got going on even when others doubt you, then your opposite action can bring stability to an unstable situation. To be patient, to wait, to stand and not grow weary of standing...sounds like a virtue that would have come in handy on guard duty, right? And to resist lying about others when you are being lied about, or not hating the haters...each of these shows the opposite reaction to a conflict situation.

Veteran Mental Health Means Remaining Grounded

If you can dream – and not make dreams your master;

If you can think – and not make thoughts your aim;

If you can meet with Triumph and Disaster

And treat those two impostors just the same;

If you can bear to hear the truth you've spoken

Twisted by knaves to make a trap for fools,

Or watch the things you gave your life to, broken,

And stoop and build 'em up with worn-out tools:

Here, Kipling talks about finding balance in our internal lives, in our thoughts and dreams. I dream a lot, think up new stuff and find myself distracted by shiny objects, but I also need to make sure that I don't get lost in the dream. And I think often, but the thinking is not enough...action must be there, too. If we get too enamored by success, or too caught up in failure, then we will not be grounded and balanced within ourselves. And which of us haven't had our words twisted to mean something that we didn't intend, or have something we worked hard for destroyed? Resisting the urge to attack the knaves or retaliate against those who destroy may be difficult, but it is ultimately beneficial.

Veteran Mental Health Means Persistent Endurance

If you can make one heap of all your winnings

And risk it on one turn of pitch-and-toss,

And lose, and start again at your beginnings

And never breathe a word about your loss;

If you can force your heart and nerve and sinew

To serve your turn long after they are gone,

And so hold on when there is nothing in you

Except the Will which says to them: 'Hold on!'

Go big or go home, Kipling tells us here. That may not seem like balance, but the response to loss certainly is. If you're going to go big, don't complain about it, because that's where avoiding despair comes in. To have the sheer force of will to keep going, even after our heart, head, back, and knees are screaming stop? A veteran knows that this is often the difference between success and failure. To hold on when there is nothing in you but sheer enduring willfulness to say hold on: that's where victory lives.

If you find yourself in conflict, take a moment. Step back and look at the situation. How can you act in an opposite manner? What can you do, like Kipling suggests, to keep cool, remain grounded, and engage in persistent endurance? If you can figure that out, then you will have followed some of his parting advice:

If you can fill the unforgiving minute

With sixty seconds' worth of distance run,

Yours is the Earth and everything that's in it,

And – which is more – you'll be a Man, my son!

Veterans, How Do You Deal with Insurmountable Obstacles?

What do you do when you are faced with inevitable or insurmountable obstacles? Veterans are go-to, get-'er-done kind of folks. I was having a conversation recently about the amount of time that it takes to get things done in the civilian world, and how we see it differently in the military. At the operational level, we move from planning to execution in a matter of months, days, or even minutes. The flexibility and the responsiveness of the military decision making process...whether it's the more formal Military Decision Making Process or the grunt level Troop Leading Procedures[8]... means that we observe something, decide how to react to it, and then react.

However, there were times in our military career, or now in our post-military lives, that no amount of effort on our part could change the outcome. When we set foot on a new duty station, we are going to inevitably leave that duty station at some point. This may be a welcome blessing...I'm looking at you, Fort Polk...but it may also be a dreaded occurrence. I've known people that spent nearly their entire career at Fort Bragg, and I sure did what I could to stay at Fort Carson for the final third of my career. There are things that we just can't influence, no matter how much we want to.

There is a concept in Dialectical Behavior Therapy called Radical Acceptance[8]. Here is a recap of what it means to radically accept reality:

- Reality is as it is (the facts about the past and the present are the facts, even if you don't like them).

- There are limitations on the future for everyone (but only realistic limitations need to be accepted).

- Everything has a cause (including events and situations that cause you pain and suffering).

- Life can be worth living even with painful events in it.

So how do we apply radical acceptance to an insurmountable problem? What do we do when we lose a loved one, despite our best efforts, or a brother or sister in arms takes their own life in spite of how we tried to save them? When the relationship ends, or the kids leave, or the job we love (or need) is taken from us? How, in other words, do we accept the fact that we can't overcome something? That we, as we describe it to ourselves, *fail*? Here are some thoughts:

These Weren't Your Obstacles to Overcome

Many of us have a challenge accepting defeat. It's the no-quit, never-say-die attitude that generates success. But how often do we place that on something that is not ours to conquer? It may not have been my fight, and I need to learn to be okay with that. It may have been an impossible fight against unbeatable odds; that does not diminish my effort, and it does not require my despair at the loss. As I've often told my Soldiers when I was in the Army, "Sometimes you have to fight the fight that you know that you just can't win, in order to be able to say that you fought." Acknowledge the effort, appreciate the struggle, but understand…it may not have been your fight from the beginning. And then, knowing that it would still end in defeat, if you had it to do all over again…still you would fight.

None of Us are Superheroes

I am not going to stop the flow of Niagara Falls. I am not going to stand against a hurricane. The forces of nature that are arrayed against me…there is no way that I am going to make any amount of impact on those things. The list of things that I can't control are numerous: the rotation of the earth, the impact of gravity, the actions of another person. These are simply *beyond my ability*. Is accepting the limitations of my ability defeat? Is recognizing that I'm not a superhero, but a fallible, mistake-prone human being, a failure? To me, it's reality. Thinking that I have the ability to beat the game when the odds are stacked so much against me, or that I can achieve the impossible, can lead to frustration and despair.

What We Call Things Changes Things

Words have power. We all know that. We can talk about euphemisms, and trying to clean up something to make it seem like something more than it's not, but a sanitation technician is still the garbage man. The Motor Transport Operator is still a truck driver. We can call it mental health, behavioral health, mental wellness, mental illness, but it's all still talking about the same thing. The difference is, the *meaning* that we place on the words.

If we call the end of the struggle against an impossible foe "defeat," then we're going to feel defeated. If we say "I failed" against something as inevitable as death, or a natural disaster, then we will feel like a failure. This isn't a "circle of life" discussion like in the Lion King, but then again, it sort of is. If we see the end as the true end, and our inability to control inevitability as a failure, then we will have truly been defeated.

Focus on What You Can Control

So how do we deal with it? Sort of like the moment after the punch, there is a moment that exists after the fight ends. What we do next, how we think next, is going to determine our future success. We are now someone who has weathered a great storm, and we may be bruised and broken, but we are wiser. There is more that we understand. We know the sting of pain in a way that we did not know it before, and, brothers and sisters, that's a valuable lesson. That's what we can control…our reactions to the inevitable. As Kipling said in "If-",

If you can force your heart and nerve and sinew

To serve your turn long after they are gone,

And so hold on when there is nothing in you

Except the Will which says to them: 'Hold on!'

I can control my will in the moment after life has knocked me down. I can control my heart, my nerves, my muscles and bones, to continue on in a life after the inevitable end. Learning that, too, is a valuable lesson.

Resistance or Resilience?

There is a lot of discussion about "resilience" in military conversations over the past fifteen years concerning our nation's conflicts. In the last few years of my Army career, I had the distinct pleasure of being able to attend the University of Pennsylvania Master Resilience Trainer (MRT) course[9] and because of this have gained a unique perspective on the topic.

Of the many definitions of resilience that are out there, the one that resonates with me is the concept that having resilience means to have the ability to recover quickly from difficulties. During our MRT course, resilience was demonstrated by the elasticity of a tennis ball, rather than the fragility of an egg. A tennis ball, when dropped, will compress, but will spring back into shape. An egg does not have the same elasticity, and instead will break if dropped from the same height as the tennis ball.

So how does this equate to life in the military?

The challenges are significant and many, including the changes that occur in the move from assignment to assignment, impending deployments, time away from home, and readjusting after deployments. Often, many of these things happen at the same time, and our families are impacted significantly. Sometimes, these things happen simultaneously. I recall, during my first deployment to Afghanistan, that my wife and I were considering our next assignment. I looked up my potential orders on the Army Personnel website and it read: 25th Infantry Division, Schofield Barracks, Hawaii. "YES! Score!" I then checked it two days later: 58th Transportation Battalion, Fort Leonard Wood, Missouri. "What? NO!" How is it possible to go from the beaches of Hawaii to the woods of southern Missouri? THAT is the rollercoaster of military assignments that we experience.

This kind of whipsaw and change requires some extreme flexibility, which develops resilience in us. When we are faced with the everyday stress of military life, there is temptation to avoid or ignore the obstacles, rather than assess and address them.

A key factor of resilience is our own mindset, our outlook on the world. Is it positive? Is it negative?

Understanding how we view the world is critical to choosing how we react to the situations in our environment. One of my UPenn instructors, Dr. Karen Reivich, said this about her own outlook:

"I'm a natural pessimist. The first thought that comes into my mind every morning is a negative one: 'today's going to suck' or 'I don't want to get moving.' I can't control that thought, but here's the key: I know that I can, and do, control every thought after that."

We have the ability to control our minds, if we are aware of them. As a matter of fact it is one of the only things that we CAN control, even if it feels out of control at the time. Viktor Frankl, in his book, *Man's Search for Meaning* one of the most influential mental health professionals of the 20th Century said this: "Forces beyond your control can take away everything you possess except one thing, your freedom to choose how you will respond to the situation." This reaction, this choice, is the difference between bouncing back or cracking.

So, when faced with an obstacle, there are three things to tell ourselves that can help with resilience, rather than resistance:

I Am Experiencing A Challenge

It is extremely important to take a step back and acknowledge that we are in the midst of a challenge, an obstacle, or a crisis. If we don't take the time to pause and understand exactly what is happening, both in the world and in our own mind, then there is a chance that we will react without thinking.

Something Caused This Challenge

"Is it something I did, or something that someone else did?" Assigning appropriate responsibility is very important. Blaming ourselves for the actions of others is not helpful, just as it is for blaming others for our own actions. Assigning appropriate responsibility is a good way of acknowledging the truth. "My car broke down." "Did I ignore routine maintenance, or did something unforeseen happen?" Deployment orders come down: "Is this unexpected, or did we know it was coming?" When assigning appropriate responsibility, it is important not to blame, but to acknowledge the appropriate role. With blame comes condemnation, negative judgment against ourselves or others. Blaming benefits no one; acknowledging and accepting appropriate responsibility benefits everyone.

I Have Control Over How I React To This Challenge

As mentioned above, we have the ability to choose how we react to any given situation. "Do I ignore the problem, avoid the problem, or rage and riot over it?" That's the path of the cracked egg. The choices that lead us to make a bad situation worse are the paths of resistance, not resilience. Instead, if we acknowledge that there is a problem, understand that there is a cause to the problem, and have the ability to provide a solution to the problem, it is much more likely that we are going to bounce back from a tough and challenging situation.

By facing these challenges with these three key concepts, we can ensure that we remain resilient in the face of pressures, rather than resistant to the changes and end up cracking under the pressure. What have you told yourself to overcome obstacles? How have you been resilient in the face of the constant change and stress that is life in the military?

PART 4

Developing Skills to Apply to Our Post-Military Life

"You can cut down a tree with a hammer, but it takes about 30 days. If you trade the hammer for an ax, you can cut it down in about 30 minutes. The difference between 30 days and 30 minutes is skills." -- Jim Rohn

You can think and talk about how you're going to react to a situation all you want, but your time is going to be wasted if you don't put it into action. Every military school I ever attended had both the classroom portion and the hands-on portion. You can study and you can learn, but if you don't apply what you've learned in real life and in real time, then you're not truly learning.

Skill development is a key part of success in post military life. There are those skills that all veterans are told that they should develop, such as resume writing, interviewing techniques, dressing for success, networking. What is often overlooked, however, are the mental and psychological skills that support these. How to handle criticism, because no two resume advice givers are going to agree on your resume. How to handle rejection, because that's going to be a part of the interviewing process. How to *want* to dress for success, *why* networking is important.

The skills discussed in the following collection focus on how to react to stressors in our post military life. How to protect ourselves when someone attacks us verbally, or when we get involved in office politics. Practical ways to respond to adversity, or that moment in the middle of the night when your eyes pop open and there's a storm in your head. Conflict is going to happen after we leave the military, just as it did when we were in the military, only it's going to take on different forms and have different impacts. How effectively we respond to conflict is a reflection of our skills.

Here's How to Protect Yourself from Attack, Psychologically

Life throws a lot of crap at you. Forget life, PEOPLE throw a lot of crap at you. It was happening when you were in the military, and it happens now that you're out. How do you protect against it?

What if you're leaving yourself vulnerable to attack?

Imagine your core sense of self as a castle tower. Inside that tower is everything that is important to you, personally. Your hopes. Your dreams. Your sense of safety. Your values. Everything that you think is right, everything you think is wrong, everything that matters to you is stored inside this tower. How do you protect it?

The Impenetrable Tower

Some veterans I work with have constructed impenetrable defenses around that tower to ensure that no one gets in. Not nobody, not nohow. There are twenty foot high steel walls, electrified, with razor wire on the top, surrounded by a moat. The moat has sharks in it, and alligators. Beyond the moat, a minefield. And more walls. Just for good measure, they randomly fire artillery from inside these defenses, lashing out at whoever happens to be around, just to make sure people know that they should keep their distance.

The Vulnerable Tower

Other veterans I've worked with leave their tower totally unprotected. No walls, no defenses, open for anyone to come along to poke at whatever the veteran wants to protect. Sometimes, this is because we're too busy trying to protect others. Trying to build the defenses for our kids, our spouses, random strangers that happen to be in need. We spend so much of ourselves in trying to help others, protect others, shield others, do for others, that we have no energy or resources to protect ourselves.

Other times, veterans will leave themselves unprotected because they've learned that there's no use in protecting yourself. It never works. Any time they start to build defenses, someone comes along and ridicules them. Mocks them for trying to protect themselves. Telling them that there's no need to do that, they have no right to do that. The veteran learns to believe that nothing they do will make a difference, so why try? A by-product of learned helplessness[1] this can lead to apathy and depression.

The Two Extremes

So the veteran leaves themselves completely vulnerable, or completely isolated. Neither of these things are very healthy; we move between allowing people to come in close and mess with our stuff, and keeping people miles away from our stuff. What if, instead, we built some logical defenses to balance that out?

Sure, set up walls. We need to protect ourselves from people who want to tear us down for their own amusement. A colleague of mine often says, "50% of the people around you don't care that you have a problem. The other 50% are glad that you do." It's an honest fact that there are people who are miserable in their own lives, and do what they can to spread that misery. Do you want someone like that close to you?

We also, however, need to be able to let people in close to us, people who matter. People we trust. Because we, humans, are social organisms, and do our best work when we work with others. It doesn't have to be an all-or-nothing proposition, however; we can decide who stands next to us, who stands a little farther away, and who we keep outside the walls.

Establishing a Balanced Defense

Consider a castle with concentric circles. You have the area close to you, next to your tower, which is surrounded by a wall. In the wall is a gate, which you control. Beyond that wall is an open space. And an outer wall beyond that. You have people in your life that you let into the inner circle…your family, your closest friends. Those who you absolutely trust. In the outer circle, you allow those who may help you, or they may not. The people in this circle may one day move through the gates to your inner circle; unfortunately, sometimes, the people in this open space once were allowed on the inside. For those who you don't know, or have shown themselves to be absolutely untrustworthy, they are put outside the castle walls.

Finding a balance between protection and vulnerability can ensure that you continue to be part of a social network while also keeping attacks at bay.

Yes, those people in your inner circle may sometimes hurt you. Dash your hopes. Dismiss your dreams. That happens, because they're human, and you're human. If it happens rarely, then you know they're good people. If it happens often, then you can evaluate whether or not you need them to be there. Realize, also, that you don't need to abuse the people in your inner circle; don't dump boiling oil on them, or set traps for them, or force them to stay in your inner circle because you need them. If that happens too much, you may find yourself outside THEIR castle walls.

Yes, sometimes you may let someone from the outer circle through the gates, and you misjudged them. It happens. They may get into your inner circle and start wreaking all kinds of destruction. At this point, acknowledge the lapse in judgment on your part, and firmly set a boundary that keeps the toxic individual from impacting your core sense of self. You don't have to go to extremes, kick everyone out of the castle, and lock yourself away. Simply disconnect yourself from that person. Come to terms with the facts of what happened. Learn the lessons that keep it from happening again in the same way.

We don't have to live according to extremes, with everyone having access to our core selves, or nobody having access to our core selves. We can choose who to let in and who to keep out. And it's that choosing, that learning, that selecting, that will bring us closer to those who we can truly trust.

Learning Self Care in Your Post Military Life

Veterans loved their time in the service. Many, if not most, look back on their time in the military fondly, if not always accurately. This comes in many different forms. One is a type of wistful nostalgia longing for the good old days when we were young, fit, and had a full head of hair. Another is an accurate remembrance of the fact that, while there may have been some times that were really, really bad, there were also sometimes that were really, really good. Regardless of how we looked back on it, sometimes we wished we were still back there.

Let's not forget, however, that we really didn't take good care of ourselves, as best as we could, while we were in.

As both a veteran and a mental health professional, I work with veterans who have left the military and, for one reason or another, are seeking mental health counseling. In my clinical training, I learned many different evidenced based therapeutic interventions; one of these is Dialectical Behavior Therapy (DBT). DBT incorporates elements of mindfulness, emotion regulation, interpersonal effectiveness, and distress tolerance.

DBT Skills come in a wide variety, but one that I teach in all of my groups follows the mnemonic, PLEASE[2]. The PLEASE skill focuses on taking care of our minds while taking care of our bodies. The mnemonic stands for:

PL: Treating PhysicaL Illness, **E**: Balanced Eating, **A**: Avoid Mood Altering Substances, **S**: Balance Sleep **E**: Get Exercise

During a recent group, one of the veterans in the group said something that I hadn't considered. "Here's the problem," he said in a frustrated voice. "This is how we SHOULD have been taking care of ourselves while we were in...and we did the opposite of almost all of it!"

He was absolutely right. The military that I served in might have preached this stuff, but service members...including myself...sure didn't practice it.

Physical Illness was Avoided Instead Of Treated

I first started to notice numbness and tingling in my right hand during Jumpmaster school, sometime in 1998. It wasn't until 2013 that I got it looked at, and was diagnosed with severe carpal tunnel syndrome. The nerve conduction study indicated that I had zero connectivity between my lower arm and my forefingers. The doc said that he had only seen that in the hands of eighty-year-olds, and another doc brought in her intern to show a classic example of uncorrected carpal tunnel that led to muscular atrophy in the heel of my hand.

In short…sick call was to be avoided, not pursued. No one wanted to be the sick call ranger. A high ankle sprain in Iraq? I went to our medic and got some crutches for a couple of days. It wasn't until seven months after I deployed that I found out that I had microtears in the ligaments. I hurt myself on a jump in 2012, which was the beginning of the end of my military career…and still had to jump twice more just to make sure I hurt myself real good.

The DBT guidance? "Take care of your body. See a doctor when necessary. Take prescribed medication." Maybe something we should put in place now that we're not in the military anymore.

Eating was Not Balanced

You know the drill. MREs, the million-calorie bombs. If you are breathing while you eat, you're eating too slow. From the very beginning, you are taught to eat quickly and cram as many calories in you as possible. Our security escort patrols typically happened in eight or ten hours segments, and when we would pull in, the first thing we would do is head to the DFAC. I would make a huge roast beef and swiss sandwich, heat it up in a microwave, and eat it as fast as I could before I fell asleep. And as great as it was at Grandma's house at Thanksgiving, going to a Stateside dining facility was enough to put the entire squadron in a tryptophan coma. We either ate too much or not enough, surviving on RipIts and Slim Jims or cramming Pogy Bait in every available nook in our rucksack.

The DBT guidance: "Don't eat too much or too little. Eat regularly and mindfully throughout the day. Stay away from foods that make you feel overly emotional." Especially as the metabolism starts to slow down, the cholesterol starts to rise, and eating too much or too little has a bigger impact on us. Probably good advice for our post-military lives.

Mood Altering Substances Were Commonplace

Do I even need to talk about this one? We all know about the typical drinking culture, which my colleague Marc Scroggins talks about in this podcast. I was eighteen years old in Germany…drinking age? What's that? I was told by someone once that there was only three things to do in Korea: go to college, go to the gym, or go to the bar. Drink, drink, drink. The use of alcohol was modeled by our buddies, our superiors, the drinking culture in the military. As a matter of fact, the Marine Corps was born in a bar…so it has that going for it. Do you think caffeine and nicotine are not mood altering substances? Try quitting them cold turkey for a bit, and see if your mood changes. It probably spiked just thinking about it.

The DBT guidance: "Stay off illicit drugs, and use alcohol in moderation (if at all)." Mornings are a heck of a lot rougher when you're 31..or 41 or 51…than they were when you were 21. Ramp it down, enjoy the day more.

Sleep? What was that?

We would pride ourselves on pulling all-nighters, both on duty and off. A good friend of mine once said, "there's nothing like dawn on the drop zone surrounded by your mates." 24 hour duty, marathon planning sessions, disrupted sleep patterns. There were times at NTC or JRTC that my 1SG had to order me to go to sleep. I was once told that my Soldiers thought I was a robot…I was always up before them and went to sleep after them, so they never actually saw me sleep. We would go and go…usually fueled by nicotine and caffeine…until we crashed. Literally; I once witnessed a crash at a stop sign in which a soldier, who had just got off 24 hour duty, fell asleep and ran into the back of the NCO he was pulling duty with. Sleep was like a unicorn…fabled to exist, but no one had actually seen it.

The DBT guidance: "Try to get 7-9 hours of sleep a night, or at least the amount of sleep that helps you feel good. Keep to a consistent sleep schedule, especially if you are having difficulty sleeping." Of course, you say to yourself, "7 hours? Yeah, right!" After 22 years in the Army, 5AM is still a late sleep for me. If I'm lucky, 6:30. I know many other veterans who don't even get that. Breaking the bad sleep habits we had while we were in can be critical to taking care of ourselves.

Exercise…We Liked It, We Loved It, We Wanted More Of It

It was always there when we were in; 5, 10, 20 mile Ruck Marches, Battalion Runs, Organizational Days. Exercise was used as a punishment…"push until I get tired, Joe"…and as a reward…"you guys get to do PT on your own today." The Airborne Shuffle is as much as an integral part of Airborne School as is hitting the ground like a sack of crap. So, out of the PLEASE guidelines, this is the one that mostly everyone in the military actually applied.

The DBT guidance: "Do some sort of exercise every day, try to build up to 20 minutes of vigorous exercise." Oddly enough, this is one habit that we DO break once we leave the service, and its probably the best one we should maintain.

How have you shifted your self-care habits after leaving the military?

Four Keys to Navigating Obstacles on a River...and in Life

People have to go through trials and tribulations to get where they at. Do your thing – continue to rock it – because obviously, God wants you here- Kendrick Lamar

Crap happens in life. We all know it. We all experience it. The irritating daily problems, like workplace jealousy or the bad boss. Traffic. Politics and the media and what the dude you went to high school with thinks about national security. Then, there's the bigger life challenges. Loss of employment. Family struggles, arguing too much, kids not doing well in school (or even talking to you). Then the huge stuff. Losing the house. The family. Death of a loved one. Involvement in the criminal justice system.

We have two ways through these challenging moments: skillfully or blindly. We may not always be aware of how to manage skillfully, but we can learn.

Being in Colorado, I've had the pleasure of being able to go on a few whitewater trips. Great fun, but the benefit is, you've got a river guide that knows how to navigate the rapids, and a bunch of other people in the boat along with you. That works as a metaphor, as far as it goes, if you have a strong support network and are seeking guidance in working through the challenges your experiencing. On the other hand, I've had the opportunity to do some individual kayaking before.

In 2008, the Army decided...in its infinite wisdom...to take a bunch of Recruiters who had gone back to the line and pull them back into Recruiting for three months. The only summer I had between two combat deployments, and I was pulled away from my family again so I could go back to the East Coast and help boost the Army's enlistment numbers. That may be a story for a whole different post, but one of the benefits was, while I was there, I was able to get some kayaking in.

I spent some time in Fredericksburg, VA, and every weekend I'd go down to the Rappahannock and rent a kayak. The Rappahannock doesn't have a lot of extreme whitewater; I think the most challenging rapids might have been Class III, and I don't pretend that I've done anything extreme, but for someone new to the sport, it's just enough of a challenge to make things interesting.

It was there that I realized that this could be a metaphor for what we experience in life. We're traveling along, things are going pretty well, it's calm. You don't need a lot of skills to move along, just being carried by the current of life. Then something changes. The water starts to run faster, choppiness starts to crop up. Rocks are just under the surface, or poking out: we're headed for rough water. We may have seen it coming, or we may not. If we've never been through this stretch, we might not know what's ahead, and it may be above our skill level to navigate. Regardless, we're in it now, so we might as well figure it out. We splash and paddle, we may try to ramp a rock in a vain attempt to just jump out of the rapids, but shortcuts rarely work. Then…it's over. There's calm river on the other side, the rapids don't last forever. They may have seemed that way, when we were in the middle of them, but they do eventually end.

So here are some thoughts on how to navigate the rapids of life, in the same way that we would navigate rapids on a river.

If You Think You Can't Do It, Then You Won't Be Able To

Our mindset controls our actions. If we think about things negatively, with absolutes, then we are going to be absolutely negative. If your thought is, "There's no way I'm going to get through this," then there really is no way that you're going to get through it. It's simply not going to happen, because you've decided, either consciously or unconsciously, that it's not going to happen. This is a typical response when faced with severe life challenges, and big rapids. "Screw this," we say to ourselves, and look for a way out. Retreat. Disengagement. Checking out. Pulling ourselves out of the river of life, stuck in a swirling side pool, maybe turning to something mind-numbing to keep us from having to figure out how to navigate. Take a look at the trials you're facing, and face them, don't avoid them. Be safe, be accurate, but most importantly, be.

Accurately Assess What Level Of Rapids You're Going Through

Looking at the descriptions of life's challenges at the beginning of this article, you can start to see where I'm going with this. The small life challenges are Class I rapids. Maybe Class II at high water. Nothing majorly challenging, not needing a whole lot of advanced skills to navigate. If we react to them as if they were Class V rapids, though, we're going to be doing it wrong. We would be living life way too aggressively. Similarly, the major life stuff are Class IV and Class V rapids…if we approach them with a mindset of they're nothing but Class II rapids, we're seriously underestimating the challenging we're going through. We could also be overestimating our abilities. If we are able to accurately understand what level of challenge we're experiencing, we're well on our way to figuring out how to navigate.

Develop The Skills You Need To Navigate Safely

If we find ourselves slamming into Class III rapids repeatedly, and doing barrel rolls and getting smashed against rocks every time, we have one of two choices: continue to get beat to crap, and develop some skills. The life skills I'm talking about here are learning how to tolerate distress. How to reduce personal suffering. How to recognize and manage our emotions. Learning how to talk to each other, how to deal with people. These skills that we develop can help us successfully navigate the rapids that come up in our life, rather than just blindly blundering through.

Recognize, And Celebrate, The Other Side Of The Rapids

When we're in the middle of life's crap, it's hard to see out the other side. We may think that there's no hope on the horizon, that this stretch of bad is going to continue for days. One of the easiest ways to calm ourselves in the midst of these challenges is to understand: there's going to be a time after this where life's not so challenging. It's going to be different, certainly. Just as we never step into the same river twice, we are never the same person after traveling along the river. We have the experience of the trials we just went through, we might have the cuts or bruises or after effects lingering. We're certainly going to have some sore muscles, if we did it right, because it was hard work. However, once we're through it, we need to take the time to recognize that it's over. Just as there was a clear beginning, there is a clear ending to this transitional messy place. If we start to anticipate the next set of rapids when we complete this last set, then we're robbing ourselves of an opportunity to learn from our experiences. Life becomes nothing but a rush from one crisis to the other, and that's not an enjoyable life…especially if we don't have the skills.

How we navigate the rapids that crop up can lead to a more pleasing, peaceful, and purposeful life. We have the choice to learn how to do it, or to not to learn how to do it. Finding someone to help us develop the skills we don't have is critical.

Four Ways to Calm the Storm Inside of You

Welcome to the middle of the night.

It's 1AM, and there's a storm inside of you. There's thunder in your heart, and lighting in your brain. Whatever happened, whenever it happened, really doesn't matter: it's the middle of a sleepless night and the storm is raging. It might be fear; for many veterans, it's anger. That's the emotional part, the thunder. The lightning is the thoughts, the rubber bullets bouncing around in your brain that are flashing and crackling.

The problem is, the storm is not beneficial. Not right now, anyway. What, really, are you going to accomplish in the middle of the night? If someone caused the storm, then guess what? They're sleeping right now. That's not meant to increase the intensity of the storm, although it might, but it's the truth. Chances are, they didn't cause the storm as much as your thoughts and rules inside your own head. Fear is the response to a threat, Anger is the response to a violation of something that's important to us. THESE are the seeds of the storm, not the words that were said or the actions that were taken. If we allow others to control our internal weather, then sunny days…and sleep filled nights…will be few and far between.

If you find yourself staring at the ceiling in the middle of the night, disturbed by the raging storm inside of you, consider these thoughts:

Let The Storm Pass

Even the most destructive hurricane, the most devastating tornado, eventually runs out of energy. The skies clear, the weather subsides. Somehow, that doesn't seem to happen to the storms inside of us, though…because we keep feeding the storm. Every time the energy of the storm starts to wane, we whip it back up again with a thought or an emotion. We're up, we're down, we toss and turn, and more lighting flashes: *great, now I can't @#%*!*# sleep. I have to work in the morning! Why is this crap still happening to me?* The more we let the lightning flash, the more the thunder is going to roll. One way to calm it…let the energy die down. This is where mindfulness comes in; if you start to notice more thoughts, the lightning flash ones, then learn how to bring about thoughts that calm yourself. Stop feeding the storm, and eventually it will go away.

Don't Try to Control The Storm

One thing that I've noticed, the more I try to control something, the less actual control I have over it. From my soldiers to my kids, from driving to exercise, the more I try to impose my will on it, the less things happen the way I want. Trying to control the storm, forcing myself to sleep, only leads to more frustration, more thunder, more lightening. Instead, if I acknowledge the storm, stop trying to fight it, and come to the awareness that there's nothing I can do about it, then there's a chance the storm will die out on it's own.

Another way that many veterans try to control the storm is to artificially suppress it. Not just with sleep meds, which may or may not be effective, but by other means. Alcohol, narcotics. Trying to control the storm by suppressing it by external means is not controlling the storm at all, but really pushing it below the surface. It's not gone, you just deactivated the part of your brain that notices it. That kind of delay doesn't solve anything, and could create more problems on the other side of artificial sleep.

Don't Let The Storm Control You

We have power over our thoughts and our emotions. We may not think we do, but it's a scientifically proven fact that we have the ability to consciously modify our thoughts and emotions. It takes work, of course, but the type of work that will lead to lasting improvement. If we let the storm control us, then we won't get anything done in the morning, because we'll be so tired that we can't think straight. If we allow the crackling power of the lighting of our thoughts to continue to streak across our mind, then we're just going to keep the storm going. Instead, going back to point one, let the storm die out on it's own.

Don't Keep the Storm Contained

I hear what you're thinking: *"easy for you to say, big guy. I bet you sleep like a baby."* Really? Guess what time it is when I write this. Well past 1AM, I can tell you that. My way of calming the storm? This right here. Getting it out of my head and into the world. Opening Pandora's Box and letting the crap out that's keeping me awake. Letting the rubber bullets of my thoughts out (in a non-destructive way) so they don't do any more damage to my sleep. Now, unless you have a really great battle buddy or a really understanding partner, you're probably not going to get a good response when you wake them up in the middle of the night for a chat. There are, however, options for you to be able to talk, if you want. Search "24/7 hotline" in any search engine and you'll get over a million results in over half a second. Not only that, I guarantee you're probably not the only one up in the middle of the night…there are online chats, I even checked my social media and have at least three contacts online right now. Getting the storm out of your head and into the world can be a key factor in letting it calm down.

How do you calm the storms inside you? Maybe you do choose the medicinal route, and it works for you. I have no problem with that, whatever works, right? Just make sure that you're not doing more harm than good. But maybe you have a way that calms you to help you sleep. A colleague of mine teaches Cognitive Behavioral Therapy for Insomnia, another connection and I were talking earlier today about mindfulness meditation, and there are likely more ways to calm the middle of the night storm. The more we know, the better we are.

In the Middle of Conflict? Try Being Radical

When I talk about being "radical," I don't mean holding on to an extreme ideological viewpoint, nor do I mean being totally awesome from an Eighties standpoint. The concept of something "radical" that I'm talking about here is based on another definition of the word: relating to or affecting the fundamental nature of something; far-reaching or thorough. An example of the word used here is, "a radical overhaul of the existing regulatory framework." In this meaning, it is complete. Total. One hundred percent, pure, unadulterated, non-contaminated being.

I once had the opportunity to attend a talk given by a leading teacher, speaker, and coach in the field of mindfulness. During his talk, the speaker kept mentioning the word "radical," in various different contexts. Radical commitment. Radical responsibility. His use of the word caused me to think: how often do we say that we're committed to something, but we're really not *fully* committed? How often do we *kind of* take responsibility for something, but not really? Perhaps, if we find ourselves in the middle of conflict, when we're twisted up inside or find our relationships twisted up outside, we may find it helpful to consider how we can apply the concept of practicing this in our own lives.

Here are five ways that I can see trying this in my own:

Radical Acceptance

This is a central concept to a therapeutic modality called Dialectical Behavior Therapy (DBT), developed by Marsha Linehan, as discussed before. The concept of mindfulness is also a central part of the modality, and a key factor that is woven throughout the entire program. Radical acceptance is actually reality acknowledgement, and is beneficial to control our thoughts, behaviors, and emotions during those times when we want reality to be something other than what it is.

According to Linehan's DBT Skills Training Manual[3] there are four things that we need to consider when practicing radical acceptance. When it comes to reality, we need to accept the following things:

- Reality is as it is (the facts about the past and the present are the facts, even if you don't like them).
- There are limitations on the future for everyone (but only realistic limitations need to be accepted).
- Everything has a cause (including events and situations that cause you pain and suffering).
- Life can be worth living even with painful events in it.

In this application, the "complete and total" concept of the word radical is applied to our accepting reality as it is, not as we want it to be. We find ourselves saying that things "should" be a certain way, when they're obviously not. Many veterans that I work with wish they were back in the military; many times, it's not possible or advisable. Wanting it to be different doesn't make it so. I have failed to practice radical acceptance many times in my life; I hurt myself on a jump in 2012. I failed to accept the fact that I was actually hurt, and jumped twice more after that (one was into a lake, so that was kind of okay). I did injury myself even more, though. Failing to radically accept reality can get us into a whole bunch of trouble.

Radical Adherence

I work with veterans who are in a veteran court program, focusing on treatment for experiences they had while they were in the service that may have contributed to their involvement in the criminal justice system. As with any program, however, each individual has a different level of adherence or compliance, and that usually catches up with them. Not practicing radical adherence to anything means saying one thing, and thinking another. "Yes, I will do what you ask" is what you say, when "Well, I do my best to do what they want" or "Yeah, I'm going to do as much as I need to" is what you think.

Radical adherence is not confined to this court program, but any type of program or relationship. I practice radical adherence to my wedding vows in my marriage. An example of not practicing radical adherence, for many service members, is when we committed to "obeying the orders of the officers appointed over me." Did we always *really* do that? I hear you, "We're obligated not to follow illegal orders, though!" That's right…but if the order was to go over to the smoking area and pick up cigarette butts, that wasn't an illegal order. It was just something you didn't want to do, and didn't feel like you should do because a) you don't smoke and b) you don't want to. It's not an "illegal" order, it was an "unfair" order in your eyes, and you "shouldn't" have to do it (see radical acceptance above). Practicing radical adherence can reduce ambiguity.

Radical Honesty

When I'm talking about radical honesty, I'm not talking about being real with someone about how that outfit looks on them, or what you really think of their haircut. I'm talking about completely and totally honest with yourself and those around you on the really important things that impact your life and those you care about. Being radically honest about the addictions in your life. Being totally, completely, and utterly committed to not lying, equivocating, beating around the bush, pulling one over on someone (or yourself), Being, or any other way of not focusing on the complete and total truth.

This is hard, especially when we apply it to ourselves. We're not completely honest about the influence that those we're hanging out with has on us. We're not honest with ourselves about the impact of the fast food we shove down our throat, our lack of exercise, or the stuff we watch on the internet. We call it "entertainment" when it's often negative, disturbing, and unproductive. Once we start practicing radical honesty with ourselves, we may find it easier to practice radical honesty with others.

Radical Joy

I can hear you now: "this 'radical' stuff doesn't sound like a whole lot of fun." It means eating healthy, cutting back on my drinking, stop fooling around, and a whole bunch of others stuff that I don't want to do (see radical adherence above). But how great do you think it would feel if we experience radical joy? Complete, 100%, unadulterated joy? We get in our own way with this sometimes. When I'm hiking in the mountains, my mind is a thousand miles away on a hundred different things. How different would it be if I just stopped and totally focused on the immediate moment and allow myself to feel complete joy? Complete gratitude?

Whatever you're doing that you enjoy. On the range. Fishing on the lake. At the gym. On the golf course. The enjoyment we feel is often contaminated by our worries about what's going to happen tomorrow, or our fears about what happened yesterday. Why allow them to get in the way? Focus on the joy you find in the moment you are in. Look at a sunrise. Listen to the wind chimes. Play with your kids. Allow yourself to feel pure, uncontaminated joy. How great would that be?

Radical Grief

And, back to the downer stuff. But is it really? When we experience loss in our lives…and we always have, and always will…the grief that we feel is often covered up by other things. Keeping busy and being task-oriented…that's the way I deal with it. *DOING* something is often easier that *FEELING* something. The challenge, however, is that the grief is always going to be there. It's part of the process of loss, and part of the process of ultimate healing.

When you feel grief, feel it. Don't deny it. Don't shove it out of the way, pretend it doesn't exist, or ignore it. Don't contaminate it with judgment, or anger, or bitterness. Just acknowledge it. Recognize it. Allow yourself to feel it, and allow it to pass, because it eventually will.

Radical focus on shared goals is something that veterans are familiar with from their time in the military.

Veterans Helping Veterans: Four Pitfalls to Watch Out For

What veteran, who still holds true to the values they learned in the service, doesn't want to help their buddy out? It's the whole concept of "battle buddy", of the Warrior Ethos: Never leave a fallen comrade. It's something that's instilled in us when we were in Basic Training or Boot Camp, reinforced throughout our time in service, and one of the major challenges for many veterans when they leave…the loss of mutually beneficial support.

Sometimes, however, that can backfire.

We saw it in the service; how many times have leaders recommend that the new guy or gal stay away from "those problem children," and were ultimately caught up in doing the wrong thing? A good troop gone bad, we say, corrupted by the lazy, mean, or disrespectful. We chalk it up to peer pressure, lament the lost potential of the "ruined Joe" and shake our heads in disappointment. A colleague of mine often says, "Half the world doesn't care that you have a problem, and the other half of the world is glad you have it." A bit cynical, perhaps, for a therapist, but there's a certain ring of truth to it.

But I'm no longer in the military, and this post isn't about peer pressure, or the negative side of pulling people down. Instead, I want to focus on the danger, and the lack of awareness of this danger, in those that are being pulled down.

I see it often with the veterans I work with and in the veteran support community I'm involved in. Once a veteran starts feeling better…they have some time in sobriety, maybe, or they're starting to manage their sense of purpose and loss, they're kicking the depression or the PTSD in the face…then they want to turn around and help other veterans. That's great; I'm all for it. The feeling that comes from helping our brothers and sisters is one of the greatest in the world. The problem is when we start to do it too soon, and the danger is slipping back down the slope of the pit that we just pulled ourselves out of.

Take a look at four thoughts on the dangers of trying to help others when we're not at a stable place ourselves.

The Easiest Way to Forget Your Problems is to Help Someone Else With Theirs

Over and over again, I see it; rather than focus on their own challenges, veterans want to turn around and help other veterans. Some of it comes from the battle buddy concept, I get it, but it can also be another form of avoidance. How do I take my mind off of my problems? Help others with theirs! Maybe this is a non-cynical aspect of the "half the world is glad you have a problem" statement: "I'm glad you have a problem so I can help you fix it." The challenge is, intent and ability are not always equal. We start out with good intentions, but overestimate our ability to solve the problems of others. We may not fully understand how we got to this semi-stable place, but since we're starting to feel good, we want to pass it on. Each of these can be dangerous to our stability.

If You're Not Stable, You Could Be Pulled Down Again

Consider this: you're climbing up a steep hill. You've gotten to a place where it's a little easier, and want to turn around and help the person behind you. If you don't take time to plant your feet, ensure that you've got a good grip on where you're at, then when you reach back to pull them up: you might find yourself back down there. You might be moving too fast. Maybe the person behind you isn't ready to get where you're at, so you reach farther…and farther…and soon, your center of gravity off balance, you tumble down. It's hard for a veteran to look back and see another veteran struggling, and not do all we can to get them to where we're at. The fact that we often forget: *you have to want it.* We had to want stability, wellness, peace in our lives to get where we're at, or else we wouldn't have gotten to where we are. We sometimes think we can put that desire, that motivation, into others, but we can't. When we try…we're in danger.

Once You're Down, You're Where You Are, Not Where You Were

The problem with tumbling back down is that you're now going to have to make the climb again. You may beat yourself up, looking up at the place you were, judging yourself. That's not helpful. Sometimes, the person you were trying to help, the one who pulled you down, is no longer around. Maybe they slid farther back down the slope. Maybe they left on their own, or some other circumstances removed them from your life. Sometimes, and this is the one that really gets you, they climbed on your back and used your shoulders to get to a place of their *own* stability…then didn't reach back down and help you up. I get it: that sucks. There's anger there. Vengeance. The problem is, we can be angry, frustrated, beat ourselves up all day long, but unless we do something about it, we're just going to be stuck. It's the truest meaning of that Buckaroo Banzai quote: *"Wherever you go, there you are."* Start where you're at. How you got there is important, but it's not important enough to keep you there.

You Got There Once, You Can Get There Again

One thing we have going for us: we know stability is possible. Wellness is possible. We may have a long way to get back to where we were, but we know it's possible to get there: we were there before. We may even still have the tools we used to get there the first time, still believe in them and know how to use them. That may make getting back up there easier. Either way, we know that stability is possible, but like I said earlier: *you have to want it. Again.* And again and again, as long and with as much effort as it takes.

Don't get me wrong, I'm not saying don't help people. By all means, help others. In my opinion, that's the surest path to peace in this life, to pass along help. The challenge is when we try to help others and we're not in a good place ourselves; we have to be aware of the danger.

Allowing our Better Angels to Overcome our Worst Demons

We are not enemies, but friends. We must not be enemies. Though passion may have strained it must not break our bonds of affection. The mystic chords of memory, stretching from every battlefield and patriot grave to every living heart and hearthstone all over this broad land, will yet swell the chorus of the Union, when again touched, as surely they will be, by the better angels of our nature. – Abraham Lincoln

My father served in Vietnam and then worked as a stockade guard at Fort Leonard Wood, Missouri until he got out of the Army in the late '60s. As you might imagine, he never talked much about either, only to say that "stuff happened'. Ironically, those stories didn't even start to come out until after I had deployed to Iraq in '06. After leaving the service, he worked in law enforcement, serving on several police forces in and around St. Louis in the early to mid '70s.

He once told me: some of the things he saw in Vietnam and in the stockade bothered him. He expected it, though; Vietnam was combat, and the stockade was the stockade. The hardest thing for him, though, was coming home to "society" and seeing the horrible things that humans can do to each other. He said, "I saw things that were as bad, and sometimes worse, than combat. The difference was, these were my people doing it to each other."

Allowing the Better Angels of Our Nature to Come Through

This isn't a commentary on the apparently worsening divisiveness that seems to be running through our country right now. Divisiveness has always been there, in one form or another; we are more aware of it now, I think, because it's in our face 24/7. Instead, this is an attempt to help others understand that we are reacting to each other, on a personal level, rather than working together to solve common problems.

The quote above, by Abraham Lincoln, were the closing remarks of his First Inaugural Address. At the time of this speech, March of 1861, seven states have seceded from the United States, and the first shots of the civil war were just over a month away. He saw it was coming, and yet he still made an effort to appeal to his fellow Americans: let us allow the better angels of our nature overcome the worst demons.

Don't Make the Other Person the Problem

I'm not talking about groups, the System, the Man, or any other conglomerate organizational group. I'm talking about our interactions with our coworkers. Our spouses and children. Systematic change comes when people with a common cause stand up for what's right, certainly, but it all starts at an interpersonal level.

I see this often with veterans that I work with. There is a conflict somewhere in their life. It usually happens with other people; their boss, their spouse. The conflict is centered on a particular problem: a situation caused by indifference. Anger. Frustration. Depression. The problem is there, for all to see, set up on a chessboard. The veteran often then takes a defensive position on one side of the problem, and the other individual takes a position on the other side of the problem. The struggle to solve the problem then begins, with both parties taking action to counteract the maneuvering of the other.

Turn Adversaries into Allies

Instead of battling against someone, reflexively taking up a posture to protect what we see as ours against someone trying to take something away from us, what would it be like to join forces to solve the problem? For both parties to come around to the same side of the chessboard, and take aim at the problem itself rather than taking aim at each other. Easier said than done, you say, and you're right. That kind of shift takes compromise. It may mean you moving to their side of the table and seeing the situation from their perspective. It may mean both of you acknowledging how it got this way, and taking personal responsibility. Never a bad thing, in my opinion.

Making that movement to a common ground to solve a common problem requires letting the better angels of our nature take over. It's the worst demons of our nature that are continuing to make the problem continue; jealousy, pride, anger, righteous indignation. Instead, allowing compassion and compromise to come through, allowing ourselves to understand the other person while maintaining our self-respect and setting healthy boundaries, can finally get to the root cause of the situation rather than making an ally an adversary.

Becoming Aware of the Better Angels of Our Nature

Again, I hear you. Easier said than done. The first thing that needs to happen is that we need to be aware that the worst demons of our nature have started to emerge. The second thing that needs to happen is that we need to become aware that we have the capacity to allow the better angels of our nature to emerge. This level of awareness is difficult to achieve in the middle of emotionally charged situations.

Once awareness is achieved, however, it's rarely enough. It's one thing to realize that there's a problem to solve; it's another thing entirely to make changes in our mind and our life to solve that problem. I don't believe that change can happen without awareness, any more than I believe that awareness alone can solve problems.

One thing I do know, however: those times in my life when I have allowed the better angels of my nature overcome my worst demons, I have been able to solve some really tricky problems. Maybe you can do so as well.

PART 5

Personal Satisfaction in our Post-Military Life

Humanism believes that the individual attains the good life by harmoniously combining personal satisfactions and continuous self-development with significant work and other activities that contribute to the welfare of the community. - Corliss Lamont

One thing I've learned after reflecting on my own military career is how personally satisfying it was. It gave me both meaning and purpose. Meaning, in that it satisfied me, and purpose, in that it accomplished a goal. Finding something in our post-military lives that gives us as much satisfaction as we had when we were in is critically important.

I find the work that I do regarding veteran mental health and wellness as satisfying as anything I did when I was in the military, and sometimes more so. It can be seen as a logical extension of leadership, caring for the troops, and all that. It is many times more than that, though. It is something that gives me meaning, because I am pleased and satisfied when I help a veteran understand more about themselves, and it gives me purpose when I see veterans making the changes in their lives to live the peaceful post-military life they desire.

The collection in this section is much more personal, sometimes shockingly so. It is an attempt to help others understand why I love my job so much, why I have such passion about the work that I do. If you can find as much satisfaction in the work that you do that I find in the work that I do, then maybe you will find that peaceful post-military life that you desire.

The Benefit of Gratitude to Veteran Mental Health

"When it comes to life the critical thing is whether you take things for granted or take them with gratitude." Gilbert K. Chesterton

How often do we spend time considering how grateful we are for things in our lives? There is a benefit to gratitude; we can be silently pleased for the blessings that we have, and we can express our gratitude in hundreds of ways. As a combat veteran, I am especially aware of the presence of gratitude, and the necessity of focusing on it in my life.

As a mental health professional, I try to help the veterans I work with express gratitude as well. This can be challenging, especially when the veteran doesn't see much to be grateful for. It is in those moments, however, that we have to *seek* out things to be grateful for. Each of us has the ability to choose how we see the world. Reality is what it is...the absence of a friend, or the presence of an enemy, these can be painful, but they can also be pleasant. If the absent friend's memory is a fond one, and if the enemy is one that we have imagined ourselves, then we can determine our own way of looking at it.

How do you seek things to be grateful for? Of course there are our families, for me that is a source of gratitude that I hope I don't take for granted, as G.K. Chesterton talks about above. And the roof over our heads, the breath in our lungs. These are all things to appreciate, be grateful for, any times and at all times. But I thought that I'd share some personal things I'm grateful for, to give you some ideas.

Babies

Man, am I grateful for babies. I love babies, and especially specific ones: the babies and little kids of my former soldiers. Those that were born, or are being born, after our deployments. I know that I didn't do anything special, but there is a peace in me that knows: this child has been brought into the world, and might not have been had I not done my part. I am grateful for the service of the parents of these children, and grateful for the trust that they had in me and my fellow leaders.

That's one thing that a veteran can look back on during their time in service. Those who you served with, your brothers and sisters: they served, and sacrificed, for the next generation. And then they came home, and began the next generation. There will be generations beyond, who will learn about what their parents did or grandparents did; the idea of that is satisfying to me. I'm grateful to have been able to be a part of their lives, and that they were a part of mine.

Sunrises

There is always a new day. The sun and the moon continue their circuit in the sky; just as a moment follows a moment, the day follows the night. This is something that Little Orphan Annie knew…there is literally another day dawning tomorrow. Regardless of how difficult life seems, there are resources to help. They're there, they're ready and waiting to support…we just have to reach out.

As I've said before, and said often, there is a moment that follows us getting punched in the gut. What we do in that moment can define us, but then there is a moment after that which we can recover from…and another, and another. While there may not always be a tomorrow, I prefer to let that end come in its natural course, and not bring it about myself. It is my solemn hope that my brothers and sisters in service feel the same.

Inspiration

A sense of curiosity is a great thing to have, and I enjoy looking at the world curiously. When I look around at things that veterans are doing, I'm inspired. Veterans are making a difference in their communities, and around the country. Through individual effort in their families, and group effort to help others, veterans are using their experiences in the military to improve their lives and the lives of others. Overcoming very real obstacles in their world, and even obstacles in their own minds, veterans are surviving and even thriving.

What inspires you to do more, to be more? To dare greatly, to make an effort in your lives and the lives of those around you? The gratitude I have for those veterans who are trying to make a difference in the lives of others is great. I'd love to hear what you're grateful for; your gratitude can inspire others. As Albert Schweitzer said,

"At times our own light goes out and is rekindled by a spark from another person. Each of us has cause to think with deep gratitude of those who have lighted the flame within us."

The Joy of the Coming of a New Day

There is something about a sunrise that I appreciate more than a sunset. A sunrise illuminates a new day of promise and potential. I've seen sunrises all around the world, and when I take the time to appreciate them, I'm able to recognize the joy of a new day.

I've seen the golden light of dawn illuminate the fog winding through Arlington National Cemetery, and have seen the sunlight march down a Bosnian hillside outside of my camp. I've seen early morning rays hit the Rocky Mountains and the Smoky Mountains, and the liquid gold roll across the North African hillside. Mornings in the Hindu Kush were always meaningful, not just for their beauty, but for the end to the dangerous night. Dawn in Baghdad sometimes provided a glimpse of the beautiful vision that it once was, rather than the war-torn reality that it has become.

None of those sunrises can compare to the sight of seeing the dawning of awareness in a veteran's eyes. When a veteran realizes why they think and feel the way that they do, and come to understand that they have the ability to make a change, it is as full of promise to me as the dawning of a new day. It's that moment when something shifts in their mind, when their awareness is heightened.

Veterans are smart, extremely smart. Soldier don't mean dumb, as an old First Sergeant of mine used to say, and today's veteran is well educated and capable of understanding extremely complex concepts and ideas. Once the dawn of awareness comes, I have experienced many veterans make some significant changes in their lives. Whatever you call it…the light bulb moment, the AHA moment, it's that moment of realization that I look for.

Often, that moment occurs, just like the rising of the sun, without any action on my part. I have found that my silence is just as powerful as my speaking, and sometimes more so, when the veteran I'm working with is on the verge of that awareness. That moment when a veteran realizes, for the first time, why they think the way they do, why they act the way they do…that's a moment that I enjoy.

The dawn of a new day is full of promise and potential, just as is the dawning of awareness. There is the promise, yet to be realized, of a life of wellness and joy ahead for the veteran. It may not happen immediately; just because the sun rises, we still have to get up and get to work. Just because awareness comes, doesn't mean change will come…that takes effort. But the *potential* for a more complete life is there; mended relationships, strength derived from growth after trauma, a lessening of the burden of shame or guilt or anger or whatever the veteran has been carrying with them for far too long.

With every dawn, darkness must come. It's inevitable, and just as the moon follows the sun. There will likely be a time in that veteran's life when awareness fades and dark falls again. The benefit of having seen the sun rise once, though, is that you can be certain that the sun will rise again. If a veteran returns to old patterns of thinking, and I'm walking with them in a dark place, then I can remind them of that time when the sun rose, and perhaps give them hope that the sun will rise again.

As a mental health counselor, I have the honor and privilege of being able to witness this dawn, this new day. Those who have served, and wish to continue to serve, are in pain, and I have been granted permission to join them in their journey to seek relief from that pain. Not to do it for them, as part of the Warrior soul is one of self-reliance, but to walk this path WITH them. To bear witness not just to that pain, but also to the joy that they had in their brotherhood, to honor their commitment and sacrifice and failures and successes. To treat them as humans, not victims, or villains, or mythic superheroes.

We have the opportunity to be encouraged by those we encourage, to be healed by those we work with to heal. This is the greatest joy that I have found in my work…to make it through the darkness in order to see the dawn.

The Wounded Healer and Veteran Mental Health

We could say, without too much exaggeration, that a good half of every treatment that probes at all deeply consists in the doctor's examining himself, for only what he can put right in himself can he hope to put right in the patient-Carl Jung

Early in my training as a clinical mental health counselor, I was having a conversation with a highly respected, and very insightful, colleague. She's a substance abuse counselor, and has significant experience in not only the substance abuse concerns of veterans, but of trauma, pain, and everything else that goes along with it. She looked at me, and said, "You're Jung's Wounded Healer!" At the time, not knowing what she was talking about, I shrugged and mumbled, "I guess." Now, with greater understanding, I see that she was right.

Carl Jung was one of the giants in the very beginning of the mental health profession, a contemporary of Freud. He is well known for his concepts of Archetypes, in that each of us has a sort of "pattern" that we follow that is in our subconscious. One of the less well-known archetypes that he described was the Wounded Healer, patterned after Chiron the Centaur in Greek Mythology. Chiron was wounded by Hercules, and although he does not die, he becomes one of the greatest healers in ancient Greece as a result of his wound.

Jung saw the pattern of wounded healer in his work as a Psychiatrist. He looked at his fellow medical professionals who were delving into the world of psychoanalysis, and noted that many of them had experienced the same challenges that they were looking to help their clients with. He noted that many of the most effective were those who had overcome their own wounds; the least effective were those who had not healed from their wounds, but tried to treat others anyway[1].

So that's what it is. Why now? Why talk about Jung's Wounded Healer on this blog? A couple of recent events brought it to mind.

Recently, another highly respected colleague and mentor shared an article that appeared in the New York Times, titled, *"A Suicide Therapist's Secret Past."* In this article[2], suicide expert Stacy Freedenthal describes her struggles with the very challenge that she is helping others overcome: suicide. She describes, in painful personal detail, her own suicide attempt during her graduate studies. She mentions in the article that this is something that she has not revealed before, and recognizes that this is her own contribution to the stigma around mental health. How can we, as mental health professionals, expect others to reveal to us their deepest secrets and fears, while we keep ours to ourselves? Stacy is a recognized expert in the field, and writes on her own blog Speaking of Suicide, and still kept this secret to herself. Until now.

A few days before this, on a trip to the bookstore (those things do exist) I picked up a copy of *Night Falls Fast: Understanding Suicide* by Kay Redfield Jamison[3]. Again, in the opening pages of the book, Jamison details her struggles with suicide and bipolar disorder, while also being a highly recognized clinical mental health expert. An author of several books, professor, and recognized expert on suicide, she is very clear about her own struggles with mental health.

"So where are you going with this," I hear you thinking. "What are you trying to say?"

I'm not about to reveal my own struggles with suicide attempts. I've never gotten to that point in my life, and hope to never do so. I've experienced it too much as a family member and friend, and have taught the warning signs in others so frequently that I am able to recognize, and react to, those warning signs in myself.

What I am going to do, however, is what Stacy Freedenthal did, and reveal my own struggles with mental health challenges in an attempt to be as transparent as possible. I often tell the veterans I work with, just as the best Preacher is a former sinner, and the best drug counselor is a former addict, those of us who have struggled with the challenges that we help others with can be effective. Hence, Jung's Wounded Healer.

This may not come as a surprise, but I have PTSD. I'll be sitting in church, or driving down the highway, and all of a sudden a memory slams into my mind. Very specific, not always traumatic, but always surprising. Random thoughts that are not connected anywhere in the present. I have control over my hypervigilance, I manage my thoughts and behaviors. It's under control, but it's always there. I have two or three "PTSD moments" throughout the year, which is great, and is a place that we can all get to.

I've also been diagnosed with Major Depressive Disorder. This, in many ways, is more challenging than PTSD. It's always kind of been there, I think, through my teenage years and leading up to deployments, but my time overseas and the PTSD certainly amplified it. The Black Dog of the Veteran Emotion[4]? Been there. The Pandora's Box of the Veteran Mind[5]? One of the reasons I have the ability to write so clearly about it is because I've lived it.

This isn't a bandwagon thing, jumping on to be able to reveal my mental health experiences like Freedenthal and Jamison. This is an attempt to let other veterans struggling with this kind of stuff to know: you're not alone. I sometimes talk about this to the veterans I work with. Not in this level of detail, but I tell them: I can't do the work that I do without having someone to talk to. I have a therapist that I see, sometimes more often, sometimes less. I can't sit with the work that I do, and contain the pain of my brothers and sisters, without having to release it through my own therapy work. My faith also sustains me, because without the strength of God I don't think I would be able to do what I do.

The thing about being a wounded healer: we need to make sure that we heal from our own wounds before we attempt to heal others. We can't hope to stop the bleeding in someone else if we're bleeding all over them as well. Do I have a handle on my PTSD and depression? Absolutely. Sometimes it creeps in, and hits me when I don't expect it. Sometimes the random thought pops up like an unwanted shadow. I know how to handle it, though, and where to go when it does.

Healing is possible.

Trust, the Key Ingredient to Veteran Mental Health

When it comes down to it, when life is on the line, we must match our deeds to our words.

After twenty-two years in the military, and another four working with veterans as a mental health counselor, I understand the immense value in the concept of trust.

Trust and the Military

The Army trusted me with millions of dollars worth of equipment. More importantly, it trusted me with the lives of some of the most amazing people I've ever known. My leadership trusted me with critical missions, impacting hundreds and sometimes thousands of service members.

Sometimes, trust is implicit in the grade or position that the military gave to us. You really expect that a leader knows what they're doing...and get really bent out of shape when you realize they don't. We trust our equipment. We trust our battle buddies. We trust the enemy, sometimes, because they can be so fricken' predictable, and at least we know where we stand with them. We trust that chow is going to show up on time, and get really bent out of shape when it doesn't, but always have a backup plan in the form of that Chicken with Noodles MRE stashed under the seat.

Sometimes, that trust is betrayed. The rates of sexual assault and harassment in the military, both on male and female, are overwhelmingly high. Those we are sworn to defend, to protect, to serve along side of, are too often those who hurt us. Just like any organization, there are those who are more self-focused than other-focused, and the leaders of the organization can do much to establish or destroy the culture of trust.

So at the end of a service member's military career, whether it be eight years or twenty years, they have learned to trust at some points and not trust at others. They have a finely tuned trust/distrust meter that served them well, or maybe not so much, but it was mostly effective.

Trust is a very fragile and powerful concept. It is easily broken, and slowly and imperfectly repaired when it is. If it stands the stress and test, however, it is a bond that binds tighter than steel.

My previous chosen profession, that of Soldier, placed immense amounts of value on the concept of trust, and prepared me accordingly. I received guidance and mentorship. I was rewarded when I cared for that trust appropriately, and punished when I abused that trust. Whether in combat or in garrison, I was more than reasonably certain that I could get the support that I needed when I asked for it.

Trust and Veteran Mental Health

My current chosen profession, however, does not have the same buy-in when it comes to a culture of trust from veterans. Like it or not, true or not, veterans think they can't trust mental health professionals.

The gift of trust that a veteran gives to a mental health professional is even more fragile and tenuous than when that service member trusted their leadership in the military. They don't know me, they don't know what I look like or am going to say to them when they sit down in my office. They know what it feels like when they think about some of the stuff that they saw or did, so of course other people are going to be horrified or nauseated or judgmental. Why wouldn't a therapist be that way too?

A lot of it, in my opinion, is that we, veterans, put more trust in misinformation than we do in reality. We accept common experiences as total truth, and react accordingly. We go from a "sometimes this happens" mindset to "this always happens." I hear it often: "They're just going to throw a bunch of pills at me." Sure that happens, sometimes. Maybe even a majority of the time, with some psychiatrists. Others that I know actually *reduce* the number of medications that veterans are on, rather than increase them. "They're not really going to understand where I'm coming from, so there's no sense in trying." There's not really any sense in that, if that's what you believe, if that's what you trust to be true.

Instead, I believe the mental health profession needs to learn how to establish trust in the veteran community. Trust that mental health counseling is a safe place for the veteran to understand more about themselves, about why they think and act the way they do. Trust that when a veteran reaches out, support will be available.

We focus so much on stigma against seeking help, and overcoming it, that we forget that when a veteran reaches out and knocks on the door, there's someone on the other side to answer. When veterans hear of that trust being broken, or betrayed, or abused in some way, then they'll share it. And other veterans will believe it, because we trust the word of one of our own before we trust the word of someone we don't know.

The other side of broken trust is hope. Hope that there is someone there who will be able to answer, who will respond when you reach out. Hope that there is a mental health professional that understands the veteran, is able to honor the trust that is placed in them, and help the veteran become more aware of their concerns.

For me, that trust is a sacred gift that I am honored to receive. And there are others like me in the mental health profession. You just have to look for us, because we're there, waiting to help.

Engage hope, and give us your trust. We'll do our best to care for it.

The Greatest Gift is the Opportunity to Help

I hold the hands of people I never touch.
I provide comfort to people I never embrace.
I watch people walk into brick walls, the same ones over and over again, and I coax them to turn around and try to walk in a different direction.
People rarely see me gladly. As a rule, I catch the residue of their despair. I see people who are broken, and people who only think they are broken. I see people who have had their faces rubbed in their failures. I see weak people wanting anesthesia and strong people who wonder what they have done to make such an enemy of fate. I am often the final pit stop people take before they crawl across the finish line that is marked: I give up.
Some people beg me to help.
Some people dare me to help.
Sometimes the beggars and the dare-ers look the same. Absolutely the same. I'm supposed to know how to tell them apart.
Some people who visit me need scar tissue to cover their wounds.
Some people who visit me need their wounds opened further, explored for signs of infection and contamination. I make those calls, too.
Some days I'm invigorated by it all. Some days I'm numbed.
Always, I'm humbled by the role of helper.
And, occasionally, I'm ambushed.

A friend forwarded me this quote the other day. It's from the book Critical Conditions, by Stephen White[6]. I haven't read it before, but once I did, I understood it. The veterans and military spouses I work with offer me their time, their thoughts, their greatest hopes and their deepest fears, and I honor it.

I watch people walk into brick walls, the same ones over and over again, and I coax them to turn around and try to walk in a different direction.

We all know Einstein's definition of insanity: doing the same thing over and over again, and expecting a different result. I see strong, proud people who seem to be tripped up by the smallest hurdle, or confused, angry people who can't understand why they've hit a dead end. My job, as a counselor, is to help veterans understand where they went wrong. To help them understand that there's a different way, a way that can help them find peace and an end to running into walls.

People rarely see me gladly. As a rule, I catch the residue of their despair.

When a veteran first comes to see me, it's because things have gone off the rails. Maybe they're told to come see me, by a family member or the courts. Maybe they've finally reached the tipping point of their frustration, and think they have nowhere else to turn. I welcome their skepticism, their mistrust, because I'm confident that I can show them a part of themselves that they don't understand. The word "catch," here, as in a vessel, I give them room in which to contain their grief. Their rage. Their misunderstanding.

I am often the final pit stop people take before they crawl across the finish line that is marked: I give up.

There are two ways out of this pit stop. The long way, which is life, and could be a life of peace and wellness, or the painful way. Painful to the veteran, painful to those who care for them. It's not a place of desperation, but a hopeful refuge, a paradox that contains all hope and no hope in the same thought. It's the fork in the road, the ultimate decision point, where we can choose to grow, maintain, or decline. It could be the delineating point in a veteran's life, when they finally stopped, set down their burden and said, "THIS is what I'm carrying. Will you help me understand?"

Some people beg me to help. Some people dare me to help.

Talking to a veteran online recently, and he said a phrase that has stuck with me: "Kicking down doors is a heck of a lot easier than talking about kicking down doors." And this came from a veteran who did both, and knows what he's talking about. For many veterans, the chip on our shoulders is the size of a city, or a valley, or an entire nation. We carry the burden of grief, of pride, of honor, and shame, and dare someone to try to figure out what to do with it. I have seen veterans who come to me in pain, but bare their teeth and growl in the same manner that a wolf would when their leg is caught in a trap. And I put up with the growl, the teeth, the anger, the pain.

Some people who visit me need scar tissue to cover their wounds. Some people who visit me need their wounds opened further, explored for signs of infection and contamination.

I've met veterans whose wounds have not been allowed to heal, years and sometimes decades after being hurt. They've not wanted them to heal, not allowed anyone to approach, until the pain and the wound become part of who they are and what they do. They long for a time before the wound, while also not wanting to give up the part of themselves that is wounded. Sometimes, the wound has healed, but healed imperfectly, like a broken bone that was improperly set. In those times, after trust and legitimacy has been established, we explore the wound together. And continue to work through the pain.

Always, I'm humbled by the role of helper.

And, occasionally, I'm ambushed.

I'm often caught by surprise at the depth and breadth of emotion. Of the seriousness of it all. I'm ambushed by my own thoughts, by the weight of the responsibility that someone has trusted me with. I am always appreciative of the fact that I, and others like me, are trusted with some of the deepest secrets of the strongest people I've ever had the honor to know.

And I'm thankful for those who have chosen to support me.

PART 6

The Importance of Remembrance Post-Military Life

"You silent tents of green, We deck with fragrant flowers; Yours has the suffering been, The memory shall be ours" — *Henry Wadsworth Longfellow.*

One of the unique aspects of military life is the impact of memories and remembrance on our post-military life. While someone who has never served in the military can be moved by the sights and sounds of battle, they will never feel the same impact as someone who understands the full measure of the impact of military service. An old soldier recognizes an old soldier, not just because of the way they carry themselves or the words they speak, but the look that comes over their face when they are in the process of remembrance.

Remembrance and nostalgia are key parts of our post-military life. We honor those who did not return from battle, the true heroes. We are, unfortunately, sometimes haunted by their memory as much as we are comforted by it, sometimes more so. For many veterans, Memorial Day or Remembrance Day does not happen just once a year, but multiple times throughout the year. Without recognizing and embracing this very real part of our post-military life, we are robbing ourselves of the richness of our experience.

These collected articles focus on the impact of remembrance, both good and bad. It is a very real part of any service member's life. It was real when we were in the military, and it is just as real now that we're no longer in it. If we get lost in the past, however, we are well and truly lost; it is the ability to visit the past, but not make our home there, which brings us peace in our post-military lives.

Memorializing the Ultimate Sacrifice

Greater love hath no man than this, that a man lay down his life for his friends — John 15:13, King James Version

Sacrifice. The ultimate, final sacrifice of one's life for others. This is the true secret that lies at the heart of military service: it is one of the few occupations where one could theoretically, and often actually, make this sacrifice. There are several others that are similar, and not all who have served would willingly make this sacrifice, but the fact is that the potential always has been and always will be there.

When I was on Recruiting duty from '03-'05, I was keenly aware of this sacrifice. Luckily, in my eyes, I was stationed outside a military installation, so the majority of the recruits I enlisted were aware of it as well. Recruiting outside a base keeps an honest person honest; there's no point in telling people BS, because they would just go back and talk to their mom, or dad, or uncle, or whoever, who would set them straight.

I once had an astute young man ask me the question: what's the worst thing about being in the Army? It wasn't a trick question, and the answer wouldn't have mattered, he was going to join anyway. He just wanted the information. I thought about it for a while; people yell at you. A lot. Sometimes there can be the senseless tasks that come with any bureaucracy. You cut a lot a grass and rake a lot of leaves. Then, it struck me:

The worst thing about the military are the Memorial Ceremonies.

The inherent danger of service in the military, regardless of combat or not, means that people will lose their lives. When a unit loses a service member, we honor them not with a funeral, but in a uniquely military fashion: with a ceremony. Unfortunately, putting these ceremonies together was something I became pretty good at. There is a right way and a wrong way to do it, just like any other military ceremony.

There are many reasons, psychologically, that we would honor those we have lost in this way. That's a key point: to honor the memory of the brother or sister. If the ceremony is held at a stateside base, much of the ceremony is for the benefit of the family, that they may know that those who served with their fallen loved one held them in great esteem. It's also a form of catharsis for the members of the unit themselves, a way to acknowledge the grief and loss in a way that may help the grieving process. My particular branch of the service, the Army, does things literally by the book, and that includes writing a book that makes sure guidelines are adhered to. In our case, it is Army Training Publication 1-05.02, *Religious Support to Funerals and Memorial Ceremonies and Service*s[1].

The publication describes the need for these services and ceremonies in this way:

Memorial ceremonies honor our fallen Soldiers and provide an environment for survivors to grieve. The opportunity to grieve can provide healing and renewal to the living that allows the unit to move forward with its collective mission.

For me, the most difficult part of the entire ceremony is the end, and in particular, the Last Roll Call. This event is so emotionally significant that the above manual recommends preparing any family members attending the ceremony for the Last Roll Call.

If you've never been in the military, let me paint a picture for you. It's 3:30 in the morning, and you and a bunch of your battle buddies are in formation. The person on charge of the whole group is checking to make sure that everyone's where they should be, which is right here along with everyone else. One of the easiest ways to do that? Respond when you hear me call your name. Private Smith? *Here.* Sergeant Jones? *Hooah.* Specialist Anderson? Anderson? Hey, where the @#%$ is Anderson? Imagine Ben Stein in Ferris Bueller's Day Off calling for someone who isn't there…only the consequences are much, much greater in the Army.

The same thing happens in a Memorial Ceremony, only for vastly different reasons. At a particular point in the ceremony, the enlisted leader in charge of the unit will go to the front of group and conduct a roll call.

Private Smith?

Here, First Sergeant.

Sergeant Jones?

Here, First Sergeant.

Specialist Anderson?

Here First Sergeant.

After each name is called, the service member stands at attention. After the last service member in formation is called, the name of the fallen is called three times. Rank, last name. Rank, first name, last name. Rank, first name, middle name, last name.

Sergeant Wolf.

silence.

Sergeant Eduviges Wolf.

silence.

Sergeant Eduviges Guadalupe Wolf.

silence.

It is in those silent moments that the reality crashes home. They will no longer answer the call, because they have answered the ultimate call. They are no longer physically in the formation, but they will always be in the formation.

They are no longer with us. They have demonstrated their ultimate love for their brothers and sisters by laying down their lives for us.

For every fallen brother or sister, those of us who remain behind feel their loss keenly. We know the sacrifice, we know the pain of loss, and we gladly bear that loss on behalf of a nation that may or may not understand. We don't have to consider this only on Memorial Day, but at any time. And at all times.

And be grateful for the sacrifice of those who have given all so that we may enjoy what we have.

The Flavor of Freedom

"For those who fight to protect it, freedom has a flavor that the protected will never know"

I often hear those words, and both agree and disagree. As a combat veteran, or even a veteran in general, I do appreciate things more now than I did before. I think that's probably true in many aspects of life; as a kid, I didn't appreciate all that my parents or my teachers did for me, and had a greater appreciation for it as I grew older.

In another sense, however, I always thought this quote was more divisive than necessary. It gave me a sense of "us against them" or sacrificial snobbery. A way to end an argument with the "you just don't understand" tactic, with me standing firm on my right perspective. I know, maybe it's just me, but there is a piece of this quote that gives me the vibe of, "look, if you haven't been there, you don't get to have an opinion about it." The quote could almost be, "for those who have never given birth, pain has a level that those who haven't will never know." Essentially, if you haven't lived it, shut the hell up.

Maybe it has to do with the word "never" and my dislike for absolutes. Really? Someone who hasn't served our country will not learn to appreciate the flavor of freedom? Like, not ever? That's hard to believe, and I imagine that it would potentially be offensive to someone who has never served but loves and respects those who have served.

However, I had the opportunity to experience something that has given me a new perspective on this quote: it has literal truth to it.

I was in San Francisco and had coffee with a colleague who lives there. Much of our discussion was focused on the need and ability for veterans to be aware of their own potential; she, in her own sphere of influence in higher education, and me in mine of veteran mental health. I was in town to present on the need for developing cultural competence in mental health counseling for veterans, and as we went our separate ways, she gave me some good advice on a place to eat, and more: The Marine Corps Memorial Hotel[2].

I don't know how often someone reading this may go to San Francisco, and stay at a hotel near Union Square, but if you do and you're a veteran, this is a must-visit.

I had dinner at the Leatherneck Steakhouse. Just me, sitting alone, looking over the San Francisco skyline. It was a great experience, and I thought to myself, "man, every service member should try this, just once. They went through a whole lot of crap, and they deserve it." The food was amazing, and yes, I had much appreciation for it.

Do you know what gave me even greater appreciation for it? Remembering my first MRE, thrown to me by a Drill Sergeant at Fort Leonard Wood in 1992. The old Cheese Omelet, in those days, was the only one wrapped in silver foil, and tasted pretty nasty. The appreciation I had for the wonderful meal that I was currently eating was enhanced by the memory of terrible meals that I had previously eaten.

That back-and-forth of present-overlaid-with-past continued throughout the meal. As I was looking at the candle on the table...yes, it was that kind of restaurant...my mind did some mental calculus through my veteran filter. Candle in a bottle...light in a bottle...fish in a bottle...Fish Boy...

In '09-'10, as I was a Platoon Sergeant for a Security Escort platoon driving through Regional Command East, Afghanistan, there was a particular village that we always used to travel through. There was a kid in this village that was there nearly every time. One day, a buddy of mine asked the truck behind him to try to see what this kid was waving at us...it was, literally, a fish stuck in an empty plastic bottle. He was waving this fish bottle at every truck that went by, grinning from ear to ear. We, of course, named him Fish Boy.

Fish Boy was a huge fan of ours. He was always the one out front, leading the other kids, ecstatic when we would come through the village. It was as if the circus came to town. He even alerted us to an upcoming attack once. Probably saved some lives, certainly made us ready for something up ahead.

As this was going through my head, I looked around at the other diners around me. They didn't know Fish Boy. My own loved ones only know Fish Boy as a story, as you do now. And they will literally never know him, as I do, because they were never there at the time.

After dinner, as I went to the lobby, I looked at the amazing displays presented. Tributes to Navy Corpsman, Korean Marines, Women Marines. The entire time, this phrase going through my mind, "Freedom has a flavor that others will never know." I stood in front of a tribute to a fallen Marine, 2LT JP Blecksmith, USMC[3].

In that display is a painting of a picture taken the day before or the day of his death, and the following tribute written by his father:

> Four weeks after the funeral for 2ndLt JP Blecksmith, USMC, a Casualty Assistance Officer delivered JP's personal effects from Iraq. A Marine Reservist Combat Photographer filmed the family

going through the belongings. Two disposable cameras were found, one of which he carried with him in Fallujah. The Marine Corps had developed the film to ensure it contained no classified material. One of the photos showed an interesting scene but the quality of the picture was poor, devoid of color or sharpness. The Photographer asked if he could take the negatives and reprocess them digitally and in high definition. The resulting photograph is what you see before you. Either it was taken November 10th or 11th, the day before or the actual day JP was killed. This photograph was one of the last visions he had of life, and shows a beautiful image. The photo is powerful, mesmerizing, and almost mystical. It depicts Marines crouching behind the wall, an air strike taking place in the distance and the beautiful sunlight shining above the clouds. You could say that hell was taking place on the ground and God was in the Heavens.

The picture, and the story, was moving to me. Impactful, meaningful. And I realized that it is meaningful to me in a way that it could not possibly be to someone who has not served in the military. Like it or not, coming out the other side of military service does give me a perspective of the sacrifice of my fallen brothers and sisters that someone who has never served does not have.

If you're a veteran, take some time to reflect on the experiences you have and how they are impacted by the experiences you had. It's part of who you are, what you know. It doesn't make you better than those who have never served, and it certainly doesn't make you worse. But it does mean you are different.

If you're not a veteran, thank you for taking the time to read and consider. It's meaningful to me, and to those veterans in your life, that you want to understand. And you will get really, really close to knowing what that flavor tastes like…and then, when you see the veteran you know looking out the window with a sad smile and a faraway look in their eye, know that the flavor of their memory is passing through their mind.

And that's okay.

Veterans, Do You Long for Days Gone By

I hear it often from veterans I talk to. I wish…

I wish I could go back to the person I was before I joined the military. Consider that for a moment…is life so terrible now that you wish you didn't exist? You don't want to not be around, but you look back at the person you were, and compare it side-by-side to the person you are now, and you would choose then. Life back then seemed so easy, you didn't have too much to worry about, your mistakes hadn't been made yet, it was long before you made a fool of yourself or screwed things up.

"That's not what I meant," I hear you say. "I don't want to go back to that age, I want to be who that person was. I want to be the happy/carefree/funny/class clown person I used to be." Here's the thing: you are. You are precisely that person, because they experienced the things you have experienced. It's not like you have shed your skin or turned into a totally different individual; you *are* that joker, all grown up.

I was thinking the other day about the level of irresponsibility that I aspired to as an eighteen year old. I initially enlisted in the Army Reserves, and spent about nine months working a dead-end job, trying out a semester of school, and realized that the best thing about that year of my life had been running around the woods with my Reserve unit. So I decided to enlist to go Active Duty. I had signed up, had a ship out date, was all ready to go. I was less than a week out from reporting for duty, when I got pulled over doing 45 in a 30.

"Do you know why I pulled you over, son?" the officer asked.

"Not certain, Officer," I said, handing him my driver's license and my reserve military ID. Even with that relatively little military experience, I thought I knew the advantage that a military ID provided in these types of situations.

"Oh, you're in the Reserves, huh?" The officer said. "Me too. I'm a First Sergeant of a company over in Illinois," He said. I thought to myself, *Jackpot.* He continued, "Yeah, sure am. And I sure as heck don't what my soldiers driving around like a bunch of maniacs! I tell them they should know better…." he proceeded to give me the good old First Sergeant butt chewing that I deserved. Everything but pulling me out of the car and putting me at Parade Rest. On top of that, I got a ticket. AND had my license taken.

So this was a Monday. I was leaving for the Army on Friday. And guess what I needed when I reported? My driver's license. And guess who didn't have two pennies to rub together? This guy. So I had to go ask my dad for the $150 bucks that it would take for me to settle the ticket and get my license back.

Do I really want to be *that* knucklehead again?

Why is it, then that we look back at the past and wish we could somehow bring that person into the present? What is it about the past that looks so much better than the present, and makes us believe that we can improve our future by pretending that the years in between didn't exist?

Looking at the Past with a Nostalgic Eye

The good old days weren't always good, as the Billy Joel song goes. We may sometimes think that life was so much easier then than it is now, but was it really? I get it, you might have been on your own, making things happen, getting stuff done. Or you might have had a good support structure, family and friends and all that. But what was so good about then that makes you want to live there? Are we looking at our past through rose-colored glasses, thinking that our glory days are behind us? That life is never going to get better than our eighteen-year-old self? Once we start to think that way, and believe it, then we're caught in a trap of longing and nostalgia. We can also look back at our time in the military, and remember the good, and forget the bad, and long for the days gone by. Is that really how we want to live our lives, looking in the rear view mirror?

Rejecting reality

I'm not a fan of denying reality. Denying reality has gotten me in trouble in the past…ignoring the fact that I'm actually hurt, and thinking I could push through the pain only injured me more. Assuming that things are a certain way, even after they're proven to be different. If we spend the time thinking about the glorious past, how much are we ignoring the actual present? Or learning how to become a better version of our current selves, rather than an idealized version of our old selves? Regardless of how we spin things or shape things, reality can't be denied. The sun comes up, the sun goes down, time continues to march on. We make mistakes, we learn, we have success, we learn. What if the happy-go-lucky guy or gal of eighteen years old was so happy-go-lucky because they didn't know stuff?

Minimizing the Benefit of Lessons Learned

I wouldn't trade a minute of today for the dreams of yesterday. I've jumped out of airplanes, engaged in combat, travelled the world. I've gotten married, raised two kids, and love the work that I do. I've had the honor of serving with some of the greatest people in the world, and mourned the loss of some of the greatest people in the world. I've screwed up, royally and repeatedly. And know what? I'm still here. Not perfect, not flawless, but

I'm here. To go back to being who I was would be to reject the lessons of life that has made me who I am.

I wouldn't trade that for all the money in the world. Would you?

PART 7

Applying Lessons Learned to Post-Military Life

There are no secrets to success. It is the result of preparation, hard work, and learning from failure. Colin Powell

Life is a great teacher. What we learn, we may not always enjoy, but if we don't learn it correctly, the lesson will be repeated over and over again until we figure it out. We can learn painful lessons by ourselves, or we can learn from others who have experienced the pain and shared the outcomes with us. That is the key to any successful military training: this is what has been tried before, this is why it doesn't work, so don't do it.

Life lessons can be shared among veterans, and often is. Balancing what is learned with our own experience gives us a greater chance of success in the future; if a trusted brother or sister says to me, "this is what I learned," I will give it more value, and then test it myself to see if what they say is accurate or not. In this way, we can take the lessons of others and apply it to ourselves.

This collection provides some lessons I learned while I was in the service, and how I apply them to my post-military life. They are different perspectives because they are generated from different experiences. They may be helpful, if you find yourself in the same situation, or they may not apply to you at all. In any event, we should never pass on the change to benefit from someone else's experience.

Four Lessons Learned from Failing Jumpmaster School

The 82nd Airborne Division Advanced Airborne School Jumpmaster Course is three weeks long. It took me six months to complete it.

There's not a leadership course that doesn't talk about the lessons that can be learned by failing. Failure lets us know that we're not perfect; it helps us understand how not to get the job done, and builds resilience and perseverance. I've screwed up more times than I can count, but when asked about one of my biggest failures in my military career, I always point back to this one. There are some pretty big lessons that I learned during that year, ones that have stuck with me nearly twenty years later.

Sometimes the best goals are the ones other people give you

I didn't want to be a Jumpmaster. Not really, not at the beginning. I wasn't alone in that, either; you didn't get any extra money for it, and Jumpmasters were always running around doing stuff while Joe put on their parachutes and then took a nap waiting to load the aircraft. No, I was in a unit with a large number of Noncommissioned Officers, but few Jumpmasters.

Our First Sergeant didn't like this. In the summer of '99, our unit had two Jumpmasters, and it was supposed to have six. So he decided to assist in obtaining the motivation to serve the unit, rather than serve ourselves.

To get into Jumpmaster School, you had the pass the Jumpmaster Pre-test. This is a written test on the proper nomenclature of every item of equipment on a parachutist's rig, and then a rigging test in which you had to properly assemble a rucksack, a release harness, and a lowering line. The written test is a 25 question test, but there were something like 150 (or more) different variations.

Our First Sergeant, in order to establish the appropriate motivation, decided to hold weekly Leadership Development courses. This was held on a Friday. After all of the Soldiers had been released. Starting at 1700 (5PM). First, we would all be in the conference room, and have to take the nomenclature test. Not the 25 question test, but the 150 question test. After that, we would be required to go outside and take the rigging portion of the test. This would typically take an hour, maybe more if a bunch of us screwed it up…which we did. Weekly.

The only way to get out of this torture was to take and pass the Jumpmaster Pre-test…we would then be excused from the weekly "leadership development." Needless to say, I started studying my butt off, and took the Pre-test just to get out of the weekly training…which, I am certain, was the whole point.

Sometimes The Goal You're Aiming For Isn't The Goal You Need

At about the same time, the 82nd had a bunch of slots to go to Pathfinder school. I may not have wanted to go to Jumpmaster school, but I sure wanted to go to that one! I had a mentor who was a Pathfinder, and it was my goal to eventually get to Fort Campbell. My uncle had been in the 101st in Vietnam, it was close to family, and…truth be told…another badge wouldn't look too shabby, either. What can I say, I was young and stupid.

So not only was my First Sergeant the master of the stick, he was also the master of the carrot. As we started getting our names in for Pathfinder school, the old man decided, "well, Pathfinder school does nothing for the company. It's good for you, but we're not going to use it. Nobody's getting on the list to go to Pathfinder School until you go to Jumpmaster School."

That was it for me. I passed the pre-test. I had the incentive to go. I might as well make it happen.

Running Into A Brick Wall Only Makes You Want It More.

When I went, and it's probably the same now, Jumpmaster school was broken up into three weeks. The classroom portion, the Jumpmaster Personnel Inspection (JMPI), portion, and the Aircraft portion. I did pretty well in the classroom portion, but…like many, many other aspiring Jumpmasters…the JMPI portion was my downfall. If I remember correctly, you had three chances to pass the test.

You have to inspect three jumpers within five minutes, not missing any deficiencies identified as "major" and only two identified as "minor." With each successive NoGo, my frustration level increased…as did my desire to succeed. After being dropped from the course, I returned to my unit, frustrated and defeated. The only question from the ever-present First Sergeant:

"When are you going back?" By that time, I was all in. Wild horses couldn't have kept me away.

Maintain Awareness, Or You Might Miss Success.

It took another cycle for me to get back into the course. I had practiced my JMPI sequence in the meantime; my First Sergeant knew some people who packed the parachutes, so we got an extra one and rigged up a Soldier so we could practice. We gave him some comp time, and I got some practice time. I hesitate to say it, but I also bent the rules a little bit by taking the parachute home with me. What are you going to do, kick me out of the Army? Like many Airborne wives, my beautiful and long-suffering wife allowed me to have her put on the parachute and practice my JMPI sequence at home!

Now I'm back. My motivation has been established, my failure smacked me in the face, and I never wanted anything more than I did at that time. When you come back into Jumpmaster school, you don't get three chances, you only get two…and I needed both.

There I was, on the edge of failure again. I remember it plain as day…I was testing in between Bragg's famous 34 Foot Towers, and when I completed my last Jumper, I stood there frozen. Did I miss anything? Did I not make it in time?

"Jumpmaster, move in front of the second Jumper," the Jumpmaster Instructor said. My heart sunk. I couldn't believe it…I was on the outer edge of my recycle window, if I wanted to come back I'd have to do the whole course again. I moved in front of the second jumper.

"Take a look at his reserve parachute, Jumpmaster," he said. At this point, I knew I was trashed. The only thing that could have been wrong with his reserve was a major malfunction, which means I missed it, and failed. "What do you see?"

I bent over, gripping the reserve parachute so hard that it, and the guy wearing it, was shaking. I'm surprised the heat from my gaze didn't make the thing catch fire. "What do you see, Jumpmaster?" He said again. Through clenched teeth, and an angry haze directed at my incompetence, I finally told him, "I don't see anything, Jumpmaster."

"That's good," he said behind me. "That's because there's nothing there. JUMPMASTER." At that moment, it hit me that he'd been calling me Jumpmaster the entire time…I'd passed!

When I stood up, I'm not sure if he thought that I was going to crack him one or give him a hug. To this day, I'm not sure which one I was going to do, either. I stood there staring at him, while the three yahoos I was inspecting were laughing at me. I probably had the goofiest grin plastered across my face, and he said, "Get out of my testing area, Jumpmaster!"

I must have set a land speed record back to the classroom. You could tell which of us made it, and which of us didn't, and I was glad to be on the side of those that had.

It was the most challenging school I attended in my 22 years, and rightfully so. It wasn't something to play with, and people's lives were literally at stake. From among any accomplishment in my career, the ability and responsibility to be a Jumpmaster is the one I'm most proud of.

And I didn't even want to do it in the beginning.

Ten Lessons from Ten Years of College

What does ten years of college have to do with veteran mental health? Maybe more than you might think. This is a story of perseverance and determination, and not quitting even though there were opportunities and urges to do so. If that's not talking about the resilience of the veteran, then I don't know what it's about.

I started this particular educational journey in 2007, standing in my company operations center in Iraq. I was already a Senior Noncommissioned Officer, and had focused on career and soldiers almost exclusively up to that point. One of my Soldiers walked in with his newly awarded Associate's Degree in his hand. He was proud, and rightly so; he asked me, "So, what's your degree in, SFC?" I told him, as politely as you can imagine, that my degree was in "none of his business." Actually, at that time in my life, I probably said more words that are unfit for publishing. My point was, I was too busy. It wasn't important to me. I didn't see the need for it. Wasn't necessary. All the stuff I told myself up to that point really didn't matter; if I didn't choose to do something then, I wouldn't be where I am today.

Fast forward ten years. Ten years is a long time to do anything, including college. Paralleling those ten years in college are four deployments, each of which I was enrolled in college at the time. Retirement. A job change. In an ever-present effort to provide value to those who spend precious time to read what I have to say, I thought I'd pass along ten lessons I learned in ten years of school.

It's Never Too Late To Start

For veterans, this is so well known that it almost doesn't bear mentioning, but it's true. When you get out after five years and start school, you feel like the old man. Whether you're 26, 46, or 56, it's never too late. The recent changes with the GI Bill mean that it's REALLY never too late. So even though I was 15 years into my military career, in the middle of a deployment to Iraq, and had no clue what I was doing, I did the only thing I could: I started.

You Never Know Where This Path Will Take You

I started off with an Associate's Degree in Counseling and Applied Psychology. It wasn't because I wanted to be a therapist. I chose it because I didn't want to get a General Education degree, and wanted the least amount of math possible. I didn't know what would happen when I started; my moment of clarity didn't come until AFTER I redeployed from Iraq.

Only after I started school did I realize what I wanted to do when I grew up…

When You're Trained In Unconventional Warfare, Everything Around You Is A Weapon

In other words, use the resources you have at your disposal. When I was on Active duty, I had Tuition Assistance to pay for my Associates, my Bachelors, my Undergrad Certificate, and a huge chunk of my first Masters. If I had waited until I got out, I would have probably only gotten a Bachelors and one Masters; instead, using GI Bill, I got two.

Never Take A 400 Level Biopsychology Course While Leading Patrols In Afghanistan

This one might sound obscure, but do your best to plan your school around your life. I found myself taking the last two major courses of my Bachelors Degree, a Senior level Biopsychology course and a Senior level Clinical Psychology Research course while deployed as a Platoon Sergeant in Afghanistan. I would be out on a security escort patrol for three days, then return to my hooch to nudge out eight page papers on the neurological impact of PTSD while my joes were kicking back playing Call of Duty. In retrospect, it was one of the most challenging parts of my educational journey. If you can do it, pick an easier class when you're in the middle of a war zone.

Know Your Limits

This one goes along with the last one. Know what you can handle, and what you can't handle. We often have trouble understanding our own capacity to handle stress, but knowing what's possible and what's not will help you tremendously. After learning the lesson above, I decided not to start my Master's Degree while deployed to Afghanistan the second time a year later. Instead, I chose an easier target, an Undergrad Certificate. Just because you can pick up heavy things doesn't mean you always have to.

Be Your Own Advocate

After finishing my Bachelors, I had a heck of a time getting the Education Center counselors to approve an Undergrad Certificate to be paid by tuition assistance. I kept getting told, "the Army already paid for your degree, they won't pay for another one." I didn't quit, and you can be sure that I researched the regulations. If you can spend hours on an interpretation of the regulation on the length of mustaches just to prove you're right about something, you can certainly spend time to do so for something that's going to benefit you. I found what I needed, took it to the counselors, and didn't take no for an answer. Advocate for yourself, in a respectful way, and you'll get what you need.

Find A Mentor

I can't tell you how beneficial it has been to find a mentor throughout each stage of the process. From the counselors at the Education Center that I talked to early in my career, to individuals who were extremely beneficial at critical junctures in my journey, none of this would have been possible if I hadn't sought out guidance. At one point, after finishing my Bachelors Degree, I was considering what I was going to do for graduate school. Here's where serendipity kicked in…I went to the Ed Center for advice.

I was sent to see a temporary counselor, because my regular counselor was not in. it turns out that the counselor I met with happened to have already walked the path I wanted to walk…he was a Senior NCO, got his Master's Degree in Clinical Mental Health Counseling, had worked in the community as a mental health provider, and was in the process of transitioning to another position on post. The 45 minutes I spent with him made all the difference in the world; had I not taken the time to seek out mentorship, and apply the guidance provided, I wouldn't be where I am today.

Do Your Homework

Of course, I mean this literally, but when choosing a degree program, do your homework. Make sure that it's going to get you where you need to be. One of the best pieces of advice that I got was from Mr. Bush, above. I was looking to choose between a three year doctorate from a non-accredited school, and a three year Master's degree. He said, "if I had two clinicians in front of me applying for a job, and one had a PsyD from a non-accredited school and the other had an MA from a CACREP school, I'd hire the MA every day of the week." Don't just jump into a graduate program to finish the graduate program; make sure it is going to get you where you need to go. If you're going to spend this much time on it, make sure it counts.

The Perfect Is The Enemy Of The Good

This has been one of my biggest challenges in college. Like many veterans, I'm a 98% kind of guy. I don't call myself a perfectionist, because I know there's no such thing as perfection, but for a long time I held myself to such a high standard that I simply didn't have an acceptable band of excellence in my life. If I aimed for 98% and hit 89%, I'd beat myself up so much that it would have been like I got a 59%. For the longest time, a "B" was unacceptable to me. I told myself that it was a failure…but it wasn't, literally or metaphorically. It's what I told myself. Do what you can, do your best, and the rest will fall in place. Don't be too hard on yourself so that you can't enjoy the fruits of your labor.

It's Always Darkest Just Before The Dawn

This has shown up so often that it's not something that I really realized. This degree is capping off ten years of college…and it almost didn't happen. Earlier this summer, with work, a significant loss in our family, and the course load I was carrying, I almost dropped it all and walked away. It was the same way with my Bachelor's Degree, as well as my first Masters Degree. I was done. I was tired of it; I wanted it to be over. All three times. I knew how to handle it this time, so even though I was on one knee two hundred yards from the finish line, I had to get up and drag myself across. What is worse…finishing the race bruised, bleeding, but standing, or realizing that the goal was in your grasp and you couldn't close the deal?

So this article was longer than normal, but so was the educational journey I just completed. Hopefully something you read here can help you on your way, maybe get you through your own educational journey. Maybe it will help you start one. I'd love to hear your story of your educational journey, so reach out. Let me know.

You never know where you're going to go, and you're not going to go anywhere if you don't start.

The Most Important Lesson about Time I Initially Ignored

One of the unique experiences of military veterans is the exposure to different cultures and nationalities around the world. If a service member takes the time to understand the culture of the people they are interacting with, they can enjoy a wide range of deep and meaningful events in their lives. The challenging thing is for veterans to be willing to take the time to understand the culture and diversity of those around them. I had an experience in my first of two deployments to Afghanistan that I did not fully appreciate at the time.

Our mission was to conduct security escort for Afghan drivers. We conducted three-day patrols, escorting supplies over a distance of nearly 120 miles. That may not seem far...the distance from Atlanta to Columbus, Georgia is only 108 miles, for example, and New York to Philadelphia is just under 100 miles. Easy to do, less than an afternoon, right? Well, try doing that while escorting fifteen to twenty vehicles that don't go faster than twenty miles an hour, over roads that are increasingly smaller, in various levels of repair (or disrepair) with a mountain on one side and a river on the other. Oh, and people want to shoot at you. A seemingly simple task takes on an entire different magnitude.

On one particular day, we were to be joined by a company of Afghan National Army (ANA) Soldiers. The way that we, the U.S. Army, did things was that we often arrive in our assembly area several hours before our required time of departure. Making sure our vehicles were ready, receiving intelligence briefings, coordinating with the units that covered the areas we would be moving through, and checking and rechecking our equipment. We had things scheduled at two hours prior, one hour prior, and time hacks every fifteen minutes after that. We had requested that our ANA counterparts arrive at the two-hours-prior mark. That didn't happen.

An hour passed. Then another twenty minutes. Finally, thirty minutes before we were supposed to leave, they pull into our assembly area.

I know what a lot of you veterans are thinking; unreliable. Unacceptable. Typical and untrustworthy. I know, because I was thinking the same thing; I had the responsibility to get this show on the road, and here these guys were pulling in ninety minutes late.

The old ANA commander and I had a short conversation, and he taught me a lesson that is valuable now, although I didn't appreciate it at the time. In my most diplomatic voice, I explained to him that he was late.

"Late for what, my friend?" He said, through our interpreter "We were told we would not be leaving for another half an hour."

"Yes, but you and your men were supposed to be here an hour and a half ago."

"But we will be driving to Asmar today, yes?" Asmar was the location of one of our Combat Outposts, about ninety miles away.

"Yes."

"And if we leave now, we will reach our destination today, yes?" I had to begrudgingly agree with him that, while true, we would not make it there in the time we had planned.

"But we will arrive today, yes?" In my recollection, he was patiently explaining this to me as if I were a child, which, to his experienced eyes, I quite possibly was. "In Afghanistan, we do what we need to. If we need to drive to Asmar today, then we will do it. It does not matter to us that it be done at a certain time, only that it be done today." He went on explaining: "In Afghanistan, we may fight this day, or we may not. If we think we will fight today, and the enemy thinks we will fight today, then we will have a battle. It matters not that the battle occur in the morning or the afternoon."

He probably didn't know it, and I didn't know it at the time, but he was accurately describing the difference between a clock-time culture and an event-time culture[1]. We both saw time differently, he and I; he saw time as tied to events, and I saw time as tied to a clock. Each of us, in our own narrow view, saw the other as "wrong," and ourselves as "right." The fact is, each of us were both right and wrong.

How much are you tied to the clock? How much do we think things "must" and "should" happen at a certain time, and in a certain way? How often, other than perhaps a random Sunday, do we take the time to consider that time is not tied to a clock, but instead is tied to an event? In his excellent book, Flow[2], Mihaly Csikszentmihalyi describes just how fluid our concept of time is, how we get "lost" in those activities that have great interest for us. Writing, perhaps, or machining, making music.

Take the time to consider how you are tied to the clock. What, really, would happen if you were more fluid and flexible? If you were more accommodating and less rigid? I'd like to hear your thoughts about it, and be more open to what life could be rather than making life be what you think it must be. It might be a good way to reduce some of the unnecessary stress we carry around.

Three Questions to Support Learning in Post-Military Life

"Change is the end result of all true learning" – Leo Buscaglia

If there's one thing that I know we did in the military, it's learn. We learned how to do stuff. We learned how not to do stuff. There is stuff that we learned that we wish we didn't know, and we learned that there was stuff that we wish we knew, but didn't. Learning happened very deliberately, like in formal schooling or planned training. It also happened more subtly, under the shade tree or the hot, hot sun.

Learning doesn't need to stop when we leave the military. On the contrary, we need to continue to learn in order to grow and be better versions of ourselves. Especially when life punches us in the face.

One of the tools that the Army used for learning was the good ol' After Action Review[3] (AAR). After many formal training events at all levels, the troops were gathered around to do an AAR. Many times, it was a "check the block" kind of thing. Someone would ask for "three ups and three downs." Three ups were three things that went right; three downs were three things that went wrong. It was supposed to be a balanced assessment, but, in true military fashion, we often talked about the "ups" for five minutes, and the "downs" for fifty minutes.

Now, I'm not saying that you have to do an AAR for everything in your post-military life. "Okay, kids, let's give mom three ups and three downs on that dinner we just ate." Probably not a way to leave your boots at the door when you leave the service. But you can learn some lessons after big events in your life. Once you leave the military. After losing a loved one. Finishing school, transitioning to a new job, anything. There are lessons in the hard times, no one knows that better than a veteran. One way to look at this is to figure out three different things: what you learned from the situation, what you learned because of the situation, and what you learned in spite of the situation.

What Did You Learn FROM the situation?

Like the AAR, there are good and bad things from any situation in life. We can identify those things that we learned from something; when we lose someone close to us, what did they teach us about life? What did we learn from them that was positive? Either they taught it to us formally, or they taught it to us by modeling it. I remember my father, a Vietnam veteran, sitting down with me before I deployed to Iraq. I was married with two kids already, and he told me, "Talk to your wife. When you get back, tell her stuff, don't keep it inside. That's one of the reasons me and your mom didn't work out." It was a lesson I learned from him, and one that I didn't (or haven't) always applied well.

Same thing from your time in service. Or from a relationship that went bad. What did you learn from it, something positive? There had to be a reason you got with that guy or gal in the first place…you learned that you liked when they did this, or there was something good that they gave you. Remembering that, focusing on that, can be beneficial for future growth.

What Did You Learn BECAUSE of the situation?

These are the three downs. These are the hard life lessons that you took away from something difficult or challenging. Maybe this might only be that you learned that you had the ability to put up with a whole lot more than you thought you could, that your capacity to endure crap was greater than you imagined. Maybe you learned that you failed to see the warning signs in yourself, or someone else, before it was too late. Anyone I've spoken to about leadership in the Army said that they learned as much from, if not more from, those bad leaders they had. We learn how to be a better leader by NOT doing what they did, a positive to their negative. What we learn because of the situation can be just as important as what we learn from the situation.

What Did You Learn IN SPITE the situation?

These are the lessons that take a little longer to understand. This is what we learn that is separate from what it was intended that we learn. Despite someone's best efforts, things just don't get through, and despite their efforts to impede your progress, you still move forward. In spite of trauma, we learn growth. In spite of rejection, we learn connection. Even though we lose someone, we can get to the point where we remember them for the life they lived, rather than the ending that came. In spite of war, we learn peace, and despite anguish, we learn joy. It's the lessons that we learn that are seemingly opposite of what we experienced; and, in that, there is growth.

What have the hard situations in life taught you? Taking the time to reflect on past hardship can drive future success, if you learn from it. We learn most from those we know, like, and trust, and if there is one group that veterans know, like, and trust, it's other veterans.

PART 8

Veteran Mental Health as Mental Wellness instead of Mental Illness

Wellness seeks more than the absence of illness; it searches for new levels of excellence. Beyond any disease-free neutral point, wellness dedicates its efforts to our total well-being - in body, mind, and spirit. - Greg Anderson

Among many of the challenges that veterans face regarding mental health, the stigma against seeking support is one of the biggest. In a U.S. congressionally mandated study about mental health services in the Department of Veterans Affairs[1], it was identified that over half of veterans who needed mental health services did not perceive that they had a need for them. Much of this lack of perception might be caused by the idea of the need for mental health services is accompanied by a connotation that this must mean the veteran is "sick" or "damaged" or "broken."

One of the ways to change the way we think and talk about veteran mental health is to focus on wellness instead of illness. Working with a mental health professional is not about curing a sickness, it's about identifying strengths to sustain while working to improve deficits.

The following collection of articles looks at veteran mental health from a perspective of building wellness instead of treating illness. It does not assume that the veteran is a broken warrior, pitiful victim, or charity cases that must be handled like fragile merchandise. It is a perspective that acknowledges that veterans are what they are: humans, fallible, and capable of amazing things in the future just as they did amazing things in the past.

Building Veteran Mental Wellness Rather than Treating Mental Illness

I don't know about you, but I don't enjoy being sick. It's no fun. Headache, congestion, pain, no energy, can't get anything done. We all have things we want to achieve in our lives, and if we're just feeling run down, it's not enjoyable. As a veteran who served 22 years in the Army, I have an extra added sense of resistance against sickness. Or maybe it's just a guy thing, as my wife says. Bottom line is, nobody likes to be sick.

If that's how we feel about being physically sick, then it comes as no surprise that we feel even more strongly about resisting the concept of mental illness. That there's something wrong with us psychologically. As a clinical mental health counselor, I work with veterans daily, and they're some of the strongest, most courageous people I know. They're also hurting and in pain.

This is a challenge that I often see when it comes to veteran mental health: many think that having to work with a mental health professional, aka "shrink," aka "going to see the wizard," aka "talking about your feelings and all that psychobabble crap," means that we are somehow weak, broken, mentally ill. With that kind of self-judgment, which is reinforced by our peers and society at large, is it any wonder that a veteran avoids talking about their mental health and wellness?

Unfortunately, it's also a subject that needs to be discussed. A mentor of mine[2] described avoiding the topic of veteran mental health is like avoiding having "the talk" with our teenagers, and silently hoping that we don't become grandparents a heck of a lot sooner than we were expecting. Perhaps, instead of focusing on veteran mental illness, we should start talking about veteran mental wellness.

Rejecting The Label Of Mental Illness

The term "mental illness" conjures up images of a padded room and a straight jacket. We think about that, then we think about *One Flew Over the Cuckoo's Nest*, and we categorically reject any thought that this label may apply to us. More importantly, for veterans, we never wanted to be "that guy" or "that gal," the sick call ranger who couldn't be counted on to pull their weight. What if we stopped looking at it as sickness or illness, though? We all know we have good days and bad days. Sometimes we're just not at the top of our game. It happens, even to elite athletes and high-powered executives.

The problem with that is, rejecting the thought that there's anything "wrong" with us may be rejecting reality. If we find ourselves working too much, drinking too much, fighting too much, or doing anything too extreme, then we're not facing facts: we're not working towards the best us we can possibly be. Does that mean we are "mentally ill?" Of course not. It means we're human.

Striving Towards Mental Wellness

Instead of looking at veteran mental health as something that is sick that needs to be fixed, how much better would it be if we looked at it as something that is working less well and making it more well? You can see the positive spin here, but it's not a "spin," it's an outlook on our life. On the world around us. If one of the rules we tell ourselves is that we only need to go see a mental health professional if we're "crazy," then we're going to reject the label and avoid getting help. If we instead change that rule to say that going to see a mental health professional means that we're focused on being the best possible us that we can be, the stigma falls away on its own.

Consider mental wellness along the same lines as physical wellness. If you step on the scale and don't like what you see, or you go to the doctor and don't like what you hear, an appropriate response is to make some changes in your life. Perhaps you get some help in doing so. Why should working towards improving your mental health be any different that improving your physical health?

We can spend all the time we want in the gym, on the road, at the range, pushing ourselves to operate at the next level. Maybe that's all the "therapy" you need. Other times, though, we might need to take things to the next level and talk to a mental health professional.

It's being able to recognize the difference between the two, and then reaching out for the support you need, that can make all the difference in the world.

Five Keys to a Successful Military Transition from Positive Psychology

Many have said it before, many times, in many ways. Leaving the military is challenging. Whether you do it on your terms, or a less desirable way, or you do it after three years or twenty-three, it's a challenge. You might do all that you can to prepare, or not be prepared at all; but the single most important thing you can do is get your mindset right. There are dozens, if not hundreds of choices you have to make: am I staying where I'm at, going home, or searching for new frontiers? Am I going to continue with what I was doing in the military, or try something new? How do I do either...or both?

As I was leaving the military, some of my colleagues were recommending logistics and supply chain management positions. While I loved the military, I certainly wasn't in love with that aspect of my service; why would I want to continue to do something I wasn't passionate about? For me, what I loved most about my "job" was getting stuff done and taking care of my Soldiers. I was able to satisfy *that* part of what I did in a new way, rather than just continuing my career in logistics.

Luckily, there are some things that you can do to make your transition successful, not in the sense that it is done easily, but that your mindset is in the right place so that the stress is lessened. Dr. Martin E. P. Seligman and his team at the Positive Psychology Center at the University of Pennsylvania[3] have identified five areas that are the foundations of resilience. The acronym to remember is PERMA[4], which stands for being Positive, having Engagement, creating stable Relationships, finding Meaning, and achieving Accomplishments. A quick internet search of the term "PERMA" will give you deeper understanding.

So how do you apply PERMA to your post-military life? Here are five areas to help you find resilience during your transition:

Finding Positivity In The Process

There are tons of ways to see the bad in things around us. For many, it's human nature to do so, but for those who are naturally pessimistic, I can all but guaranteed that you have a friend or family member who's willing to help you point out the negative. See my example in the beginning of this post: *Logistics? Really?* I was good at it, and it's okay, as far as it goes, but it would be hard for me to put a positive spin on it. But finding something positive to counteract the negative is important if you want to lower the stress of an upcoming disruption. Even seeing it as a disruption would reduce the positivity. Instead, try to see it as growth, change, an adventure, or forward progress in your life.

Developing Engagement

Engagement is being fully involved in the task; so involved that you might lose track of time. It happens when everything seems to be going exactly right, flowing along, where you're on track so much that the job takes care of itself. This is also known as the concept of *flow*, where we get so caught up in doing what we're doing that it all becomes too easy. One of the recommendations that Dr. Seligman talks about to achieve engagement in undesirable tasks…like planning your transition…is to determine what your signature strengths are, and then do those tasks using your signature strengths[5]. For example, if one of your strengths is gratitude, show gratitude and appreciation for others, and then develop a sense of gratitude towards what is happening. Be deliberate in showing gratitude as you go through the tasks of a transition, and show appreciation to those who supported you as a family. Or if your signature strength is humor, try to deliberately see humor in the many crazy things that happen during a transition. By engaging your signature strength, it's more likely that you will find engagement…and ease…in getting through task of leaving the military.

Sustaining And Creating Relationships

Another key aspect to developing resilience is to have healthy relationships. This is critical with service members, spouses, and children. Each relationship must be considered and sustained during this critical time. One of my son's earliest memories was during a PCS from Maryland to Colorado, of him eating pizza and watching *Blue's Clues* in his room while the movers were taking boxes. Most of all, he remembers the attention and care we were giving him, and it wasn't a scary or bad memory.

Developing relationships during your transition is also important. This is what we call "networking," but not the kind of networking that means throwing out business cards like bullets from a business card gun. It means developing real and meaningful relationships, having actual conversations with others. There are hundreds of meetups, groups, events for veterans all over the country. Technology allows us to start communicating with people at the new location, learning what we can, not just from websites, but from people. Establishing new relationships, while sustaining old ones, can positively impact your transition.

Understanding Meaning

Dr. Seligman indicates that we find meaning in being connected to something bigger than ourselves. In my work as a mental health professional, this is one of the most significant aspects of post-military life that many veterans feel the loss of. They are no longer connected to something bigger than themselves, and it no longer has meaning for them. Remembering that we are part of a larger family, a family that looks out for and supports each other, can go a long way to establishing meaning in our lives. Remembering that the support that we provide each other, individually, makes our nation stronger as a whole. Being aware of the meaning, the importance, of the support we provide, can help us reduce the stress of the move all together.

Recognizing Achievement

When have you not gotten through a difficult patch, and looking back, found a sense of satisfaction at a job well done? Recognizing and celebrating the achievements we accomplish is a great way to find more positivity in our lives. It's not just the achievement of getting it all over with, but a way to recognize achievement throughout the entire process. As my friend and colleague Jeff Adamec is often fond of saying, "You have to celebrate the little victories." Got the kids pre-registered for their new school? Celebrate! Checked another task off of the list? Score! We need to recognize the individual accomplishments, and celebrate the wins we have as a part of a larger goal. Even the accomplishment of getting through a stressful day without flipping a lid is an achievement, and could be treated as such. In this way, we build our resilience, our positivity, and our defense against stress.

The more we look for these things in the stressful transitions in our lives, the more we will be able to use them. When things go wrong...and they will...how we react to them is more important than the event itself. By applying PERMA to our transition, we are setting ourselves up for more success than failure.

Three Reasons to See a Mental Health Counselor...And Its Not What You Think

You don't have to be in crisis to talk to a mental health professional. As a clinical mental health counselor, I often see veterans when life has spun out of control...the drinking has gotten too much, they're about to lose their job, their family...their life. Metaphorically, if I were an emergency room doc, some of the men and women I see would be walking in with three fractures, bleeding from a thousand cuts. The national organization, Mental Health America, says it well: why do we wait for stage 4 mental health conditions to take action[6]?

Before Stage 4 is an idea that takes the concept of taking care of our physical health at early warning signs and applies it to our mental health. This is from their website, and says it better than I can:

> When we think about cancer, heart disease, or diabetes, we don't wait years to treat them. We start before Stage 4—we begin with prevention. When people are in the first stage of those diseases and are beginning to show signs or symptoms like a persistent cough, high blood pressure, or high blood sugar, we try immediately to reverse these symptoms. We don't ignore them. In fact, we develop a plan of action to reverse and sometimes stop the progression of the disease. So why don't we do the same for individuals who are dealing with potentially serious mental illness?

So why would someone seek out a mental health professional if they weren't crazy? I hear the objections. I get it. "Hey, employers already think all veterans have PTSD, why play into the stereotype?" That might be perpetuating the stigma, not reducing it. But why wait until you're in crisis before you do something about it? So here are three reasons to find a good mental health professional to talk to.

To Become A Better Version Of Yourself

You may be doing pretty well with life. The transition isn't going too bad; you've got a job. Granted, it may not be as cool, or as satisfying, as when you were in the military, but what can you expect? At least it pays the bills, right? Or maybe you're in school, and it's going okay. It's not what you thought it would be, maybe a little easier, maybe a little harder. Just because you're doing okay, however, doesn't mean you can't be doing better. I've worked with veterans who were doing okay, but didn't realize that there were steps they could take to improve. An okay life is not a great life, so why settle for a little when you can get a lot? Mental health professionals have the training, education, and experience to be able to help you get insight into how to make a good life better. You might go to a personal trainer to take your fitness to the next level, and athletes have coaches to help them develop. Why don't we do the same for our mental health?

You Did It To Your Gear While You Were In The Military

Nearly every piece of equipment in the military had a preventive maintenance schedule. Maybe you didn't always follow it, but when you didn't and you needed something, it always broke. Preventive Maintenance Checks and Services (PMCS) is what we did for our vehicles, our weapons, our communication systems, and our gas masks, so why not do preventive maintenance on our way of thinking? Taking a look at our actions and behaviors, figure out what's working, what's not, and doing something about it? If, while we were in the military, we always worked on developing our strengths while minimizing our weaknesses, why do we avoid doing the same when it comes to our thought processes and personal interaction with others? Let's kick the tires and change the oil, and catch stuff while it's small. Before it becomes catastrophic.

You Saw And Did Things Outside The Normal Realm Of Human Existence

I'm not talking about combat, although that's there, too. But the things we did when we were in the military aren't things that people do in the normal course of their daily lives. I was talking to a group of veterans, and I asked them if they missed the gas chamber. For those of you reading this who aren't veterans, part of basic military training is to be exposed to tear gas, or (as its known in the military) CS gas. We jumped out of perfectly good airplanes and jumped off of tall buildings with ropes tied around our waists. We walked for miles and miles with heavy weight on our backs and ended up where we started. Why? In order to develop the physical strength and endurance to be able to fight and win our nation's wars. Years of doing that changes you, often in ways that you may not be aware of. It didn't mean you were crazy then, and doesn't mean you're crazy now…but trying to make sense of why you did it, and how it's changed how you see the world, isn't crazy either.

There's probably more reasons than this. I might be writing a pretty big blank check for you to cash in with your local mental health counselor, but they're professionals, they can handle it. Of course, going in and checking it out will take us overcoming some of the stereotypes and preconceived notions we have about therapy and counseling. It's not all about getting meds thrown at you, or someone sitting cross-legged on the floor burning incense. There are mental health providers out there that "get it," and you can find them. You'll never know if you don't check it out.

Growth, Maintenance, or Decline: Daily Choices that Impact our Future

As I travelled along my journey, and paused, I came to realize that I was standing in front of three mountain paths.

The journey so far had been easy at times, and hard at others. Some of the paths that I had walked looked easy and became hard, and others I had walked had looked hard and become easy. The farther I went, however, the more awareness I had of my choice of paths.

At the beginning of my journey, my choices were not deliberate. I allowed myself to be led, or sometimes I blindly chose the path that took me to an unknown destination. Sometimes I would read the signposts indicating what lies ahead, and believe them, not knowing that the signposts were there for other people, and not me.

At this moment, however, I stopped, aware of the three paths before me. The path on my left climbed upward. It looked challenging, but I knew from past experience that the rewards at the end would be equal to my effort. I had chosen the upward path before, and at the end I found strength, and amazing views, and a newfound sense of purpose and meaning.

The path in front of me continued straight ahead, bending around the curve of the hill to an unknown destination. It was a familiar path, a maintaining path, and one that I knew that I could safely take if I thought that the challenge of the upward path was too great for me at this moment. Again, from my experience, knowing that my overestimation of my abilities has caused me to fail in the past, the middle way could be not only the choice of safety, but of preservation.

The path to the right...that one leads downward. It looks to be the easiest of the three, but I know that the rock and scree and scrabble could make it hard to climb again. This is the path that I had blindly taken too many times before, the descent into unknown and has the potential to wander aimlessly in a dark place, below the sun.

This time, however, I have a choice to make, and I am aware of that choice. The journey so far had not been easy, but it had been my journey, and one that I cannot start over, no matter how much I want to. It is this point right here that I have the ability to impact.

How aware are we of the choices we make in our lives? How does it happen that we take the time to pause, and consider, before moving forward? The metaphorical decision above is one that we consistently make, even if we don't know it. I've written before about Posttraumatic Growth[7], the idea that we can survive, and even thrive, after experiencing a challenging life event. Consider me, or yourself, at that decision point described above. We're not just pausing; we're looking up from the ground. We're shaking our head to clear the cobwebs after the hit, or the blast. We're wiping the sweat from our eyes, taking heavy, ragged breaths, looking around to make sure everyone's all right. Or we're recovering from the shock of what we've just seen, breath caught in our throat, heart rate increasing. Now what? And then now what, when we think back on that moment, after waking from a nightmare or reacting to something that reminds us of that?

These choices in our lives are not two-dimensional, not linear. They are 3D, 360 degree choices. We are moving forward on this path, unable to start over or turn around. We are only moving forward, and making the choices in front of us: growth, maintenance, or decline. We can make the choice of which path we take; we just need to be aware of it.

I picked up my rucksack, and turned to the left, uphill. I choose to grow, as difficult as that may be...

The Difference Between Meaning and Purpose in Veteran Mental Health

Meaning and Purpose. Purpose and Meaning. It's the Holy Grail of post-military life: searching for something that will satisfy the need for meaning and purpose that a service member had while they were in the service. It's more than just something to *do,* it's a desire to find something to *be.* I've written about the need for finding meaning and purpose in our post military lives, as have many, many others. It's one of the aspects of veteran mental health that go beyond just PTSD and TBI, and into places that medications can't touch, coming to terms with the past can't alleviate, and removing depression or anxiety can't satisfy. With the number of veterans I work with as a mental health counselor, this is the single most common difficulty that veterans are looking to overcome.

But what the heck does it mean?

What does "meaning and purpose" refer to? Are they the same thing? Or different? If you say a word to yourself over and over again, it kind of means nothing, right? Sometimes "meaning" and "purpose" are interchangeable, and sometimes they're very different. Meaning can be something that is important to you. My dogs are meaningful to me, my work is meaningful to me, life and veterans and freedom are all meaningful to me. I value them, I appreciate them, they satisfy me. Purpose can also be something that is important to you: I find purpose in the work that I do, the writing I create, the effort I make to support my family. I value them also, and appreciate them, and they satisfy me. And there's a clue to how I see meaning and purpose differently, and the idea I'd like to get across today:

Meaning is value that we place on things from an internal perspective, and purpose is value that we receive from an external perspective. Meaning is the internal push, while purpose is the external pull.

We put them both together because we need both in our lives to be fully satisfied.

Walk with me a bit, and I'll break it down a little further:

Purpose without Meaning Doesn't Last

Considering purpose as something that brings us value from an external source, it is simply: something to do. A task to be accomplished. If I have something to do...write this article, see a veteran client for therapy, write a report...I have a purpose. I have a task to accomplish that will bring about a result. This is "work," in a sense, something productive that I must do. It is possible to have purpose without meaning. Ever had a dead-end job, where the only thing it does is put money in your pocket? You didn't enjoy it, it wasn't satisfying to you, but the only reason you did it was because you had to. That's purpose without meaning…and as soon as you find something better, you get out of there.

We had stuff in the military that gave us purpose, but didn't mean anything to us. Getting smoked, anyone? The task to be accomplished…"do flutter kicks until I get tired"…was there, but it wasn't meaningful. It wasn't something that we continued to do even when we didn't want to. Or guard duty? Mind numbing hours staring at the same barren hillside, just in case the enemy horde suddenly materialized on the other side? Again, a task to be accomplished, but it didn't have meaning, an internal drive or satisfaction.

Meaning without Purpose Is Not Satisfying

Just like you can have purpose without meaning, you can also have something that satisfies you without actually having a purpose. Six hours of Call of Duty, anyone? Or binge watching Netflix. Sure, you're entertained, you're satisfied, but nothing was actually accomplished. We don't *do* anything. It's like soda: empty calories. We may find something in our post-military lives that we enjoy doing, but if we don't have the external pull of a purpose, the meaning doesn't go anywhere.

We can find something meaningful but not give us a sense of purpose if the job is too simple. Like my example of the dead-end job earlier: if it's assembly line work, or sweeping up sawdust in a lumber mill, it's not challenging enough for us to keep our attention. Sure, it satisfies us to the point of meeting a need, but then what? Or, if we're stay-at-home parents. The kids are pretty good kids, taking care of them is meaningful to us, but it's on auto-pilot and not that hard. If the purpose is not challenging enough for us, then it's not satisfying.

Meaning and Purpose Together are the Key to Post-Military Satisfaction

So finding something that is both meaningful to you and purposeful for you is critical for satisfaction in your post military life. Finding something you enjoy (meaning) that also gives you a sense of accomplishment (purpose) is key to finding balance. They don't have to come from the same source, either. You can satisfy a sense of purpose in your life by working at your chosen profession, and find a sense of meaning by going home and growing a garden. In his book *Flow: The Psychology of Optimal Experience,* Mihaly Csikszentmihalyi[8] describes a guy who builds railroad cars for a living then

125

goes home and works on his rock garden, which includes sprinklers and lights to create rainbows and shadows. A thing of beauty to look at, which he finds meaningful.

So meaning as an internal motivator, and purpose as an external motivator. We must find both in our post-military lives to truly be satisfied, and continue to make as big a difference in our communities as possible.

Veterans, Use Your Skills to be Indestructible

Many of the veterans I work with can get frustrated during their post-military lives. They were once kings and queens, warriors and poets, master and mistress of all they surveyed. Okay, so maybe they weren't in charge of all that, but they had a heck of a lot of responsibility for some really important stuff at a young age.

Moving from a familiar environment to an unfamiliar environment is always challenging. Consider the move to a new duty station, a different unit, even moving from one sector to another during deployment: you can get all the information you want, do all the research you need, but still you don't *know* until you have boots on ground. We were able to make ourselves indestructible then, why is it challenging after we leave the service?

I'm guilty of feeling this frustration as well. I have a message that I want to spread…veteran mental health and wellness is critical to a successful transition…and sometimes it's hard to get above the noise. Hundreds of people are talking about veteran mental health, veteran suicide, PTSD, veteran mental health and wellness…but where are the voices of those who are perhaps most qualified to speak on the topic, the mental health professional? I know I'm not alone in this. A colleague refers to it as feeling like he's yelling in an open room; I often talk about shouting into the fog.

So what keeps me going? Why continue in spite of the frustration and frequent lack of response? Because I don't have quit in me, and you probably don't have it in you. I want to be the John McLane my chosen space: the fly in the ointment, as he said in Die Hard. The monkey in the wrench, the pain in the ass, if that's what's needed. I want to be as indestructible as a cockroach, as long lasting as a freakin' twinkie. When the frustration, and even doubt, starts to creep in, I rely on the lessons I learned during my military service.

Tap Into Your Warrior Ethos

Every branch of service has it's creeds and mottos. The unwavering resolve that we had while we were in the military can be applied to your post-military life as well. As an Army veteran, I am most familiar with the Warrior Ethos[9]: *I will always place the mission first. I will never accept defeat. I will never quit. I will never leave a fallen comrade.* That mission, for you, is whatever you choose it to be. For me, it's bringing awareness to veteran mental health; for you it may be getting a job. Or overcoming addiction. Or being best whatever it is you're doing that you can possibly be. That never quit, maximum effort attitude that you possessed when you were in the military didn't go away…it's still there. You just have to pull it out of the back of the closet, where it's sitting next to your old PT uniform, and put it to work.

Recognize the Strength You Possess

Whatever you did in the military, chances are you were successful. Marksmanship, physical fitness, proficient in your tasks and drills. You got stuff done while you were in the service. Maybe, like me, you found yourself in an occupational specialty that you weren't really all that thrilled about; so perhaps, also like me, you found ways to stretch beyond your capabilities by pursuing challenging assignments. Jumping out of airplanes or helicopters. I get it, not all veterans are Rambo and GI Jane, but chances are you had success while you were in the service. YOU know it; all you have to do is remember it, and believe that you're going to be just as successful outside the service as you were in it. It took hard work for you to succeed when you were in, why would it be different now that you're out? Remembering past success, and how you got there, can lead to future success.

Recognize Your Own Value

One thing I see often with transitioning service members is that we take the first job that will accept us. I know I did; I enjoyed it, it kept me working with veterans, but it really wasn't what I *wanted* to do. Recognize the value that you have, both as an individual and because of the skills you gained while you were in the service. You are worth more than you think you are; you may know the value of the problem solving, determination, never-quit attitude and skills that you have, but maybe your prospective employer doesn't. If they don't see it: their loss. If they see it, and offer you a position, but it may be less than what you think you can handle, then take it. If working in the mail room (if that's even a thing anymore) pays the bills, then go for it...and your value will quickly be identified, and you won't be in the mail room for long. Just as you were given more responsibility when you demonstrated competence when you were in, the same thing will happen when you're out.

Bottom line: don't let the obstacles in your way become a dead end. They're not insurmountable, they're just in the way. Overcome the insurmountable by being indestructible...then turn around and help your brother or sister in arms.

Mentorship and Veteran Mental Health

mentor (n) 1. a wise and trusted counselor or teacher. 2. an influential senior sponsor or supporter.

From the first moment that a service member steps into an office or motor pool at a new duty station, they look around for someone to let them know what's going on. If they're smart, they sit back and observe things for a while before they start running their mouth or wrecking stuff. They look for those that have been there longer than they have, to show them the ropes, to let them know how the culture is, to help them get the lay of the land.

The first thing they look for is a mentor, even if they don't call it that.

A significant aspect of military service is the personal and professional development that goes on in each individual. We had someone that we looked up to; a senior platoon member, our squad leader, a hard core platoon leader. That commander that was tough, but fair, and looked out for their troops. Sometimes, however, service members chose their mentors poorly; they would fall in with the "wrong crowd" and learn the wrong kind of ropes. Good or bad, the examples of the leaders we had while we were in shaped us, defined us, made us who we were as Soldiers or Airmen or Sailors or Marines.

If we were motivated to succeed in the military, we sought out those individuals who were knowledgeable and effective. In the Army, we called it being "technically and tactically proficient." If we wanted to do more, get better, be stronger, we looked for those who knew how to get us there and had the willingness to share their knowledge.

That led to success in the military, and I'm here to tell you: working with a mental health professional can do the same. You don't have to be crazy, have PTSD, be challenged by severe depression, disrupted relationships, or substance abuse. Transition from the service to your post-military life has significant challenges that are not related to a diagnosable mental health condition, such as a shift in identity. A loss of a sense of purpose and meaning. Trying to figure out how to meet old needs in new ways. If you looked around for someone who was technically and tactically proficient in your chosen endeavor when you were in the military, why not do the same once you got out? Because people would think you're weak, you don't know what you're doing? I don't know about you, but that kind of thinking has caused me a lot of pain in the past.

Military Mentorship

Since my retirement, I've thought a lot about the mentors I had over my 22 year career. My first squad leader taught me how to be a Soldier. When I was newly promoted and reported to the 82nd Airborne Division, my first squad leader there taught me how to be a Noncommissioned Officer. I recall a Platoon Sergeant who pulled me and another squad leader aside, took us over next to a tree, and asked us: "Have you ever done a sector sketch?" For those not familiar, a sector sketch is a drawing that shows where each of your unit's positions were, what direction they would be firing in, main terrain features, that kind of stuff. Jeff, the other squad leader, and I hadn't had anyone take the time to show us how to do one. Our Platoon Sergeant didn't have to teach us...he could have knocked it out himself, turned it in, and went to sleep. Instead, he took the time to teach us something we didn't know, in order to improve our tactical proficiency. All down the line of my military career, I can point to someone at each level of my leadership that helped mold and shape me, for better or worse, into who I was. Mentorship was such an ingrained part of the military lifestyle that it came easily and naturally to us. When we got out, though, did that change?

Civilian Mentorship

My civilian career as a mental health counselor overlapped my military career. More specifically, the two careers were bridged by my educational pursuits, and I needed mentors in both my educational pursuits as well as my current profession. So I looked for them. Sometimes they cropped up unexpectedly; I just so happened to go to the Education Center to talk about choosing a Master's Degree program, and the counselor I talked to turned out to be just the right one I needed. He was a retired Army Noncommissioned Officer, had gotten his degree in clinical mental health counseling, and was preparing to transition to a position with the Army Substance Abuse program. He was at the education center for less than a month, and just so happened to be there on the day I walked in; but I credit that forty-five minute conversation with him as a significant turning point in my educational career. I would have made some serious mistakes in choosing programs, if it weren't for him. Later in my development, I was blessed enough to be connected to another mentor, Dannette Patterson[10], who has supported me...and prodded me...in my new career.

Qualities of a mentor

In my opinion, a mentor is someone who has your best interests at heart. Who gives you the advice you need, rather than the advice you want. Ideally, a mentor is someone that has no interest outside of your success. Your workplace supervisor may be a great teacher, and a mentor, but they have a vested interest in the success of your organization as well as your personal success. I had opportunities in the military which would have been great for me to pursue, but it would have meant that I would have to be allowed to leave the organization I was in. Guess what happened? I stayed put. A true mentor is one that helps you understand your needs better, and generates awareness in you of what is keeping you going and what is holding you back.

Mental Health Professional as a Mentor

Mental health counselors, therapists, psychologists, whatever you want to call them, they can be seen as mentors, too. We don't give advice, or at least, I don't. I don't tell the veterans I work with what to do. I don't judge them if they make a choice that's not in their best interest...I support them. I have the technical knowledge of skills and techniques that can help them see things in a different way. I have the ability to help them apply that knowledge to their situation. I, and other mental health professionals, have the ability to support service members in their transition and help them change how they think about and see the world, in order to be a better version of their excellent selves.

If you're struggling with a resume, you reach out to organizations like Hire Heroes USA to help you write it. If you're struggling with interviewing, you find someone to help coach you through that. If you're not sure how to improve in your chosen profession, you look around for someone to help. Why don't we do that when we're struggling with stuck, negative, defeated thoughts? When we start drinking a little too much, or find that we're arguing with our spouse more than we used to? When we find ourselves getting frustrated at all the little things? The resume, the interview, the professional development, all of those are structures that are built on the base of our mental health and wellness, and finding a good mental health professional can help us make sure that base is steady and secure.

Veteran Mental Health is Focusing on the Future, Not the Past

I hear you all ready: "Get out of here with that 'therapy is about the future, not the past' crap. Talking about my past is what therapy is all *about*." Granted, when a veteran sits down to talk to a mental health professional, they are going to talk about the past. The therapist is going to inquire about it, and the veteran is going to answer. But is that any different than going to any other health professional? If I go into the doctor with a broken arm, maybe they don't need to know *how* it happened. It is definitely something that is in the past that will be addressed, however. Identifying the causes of the current distress or behavior is sort of necessary in order to move forward with a plan to manage the distress or change something in our behavior.

Just because the past is something that is talked about in therapy, however, doesn't mean that's what it's *about*.

The idea that mental health counseling is about the future goes along with the idea of looking at it as mental wellness instead of mental illness. Of seeing the process of talking to a therapist as a sign of strength, and wisdom in applying the appropriate measures in a difficult time, rather than weakness. I mean, why else would you go to see a doctor about a broken arm, unless you want that doctor to use their expertise to a) stop the pain (a pain-free future), and b) set the break so that it heals properly (a future of full functionality of your arm). The same can be applied to working with a mental health professional.

A Future of Peace

I've said it often; for veterans, our wars are over. Whether we are twenty-seven or seventy-seven, our time of service is in the past. There was some cool stuff that we did, and horrible stuff that we did, but it's behind us now. In our post-military lives, we're not kicking down doors or running ammo or calling for fire on a target. We can use the skills that we learned in the military to lead productive, peaceful, meaningful post-military lives.

Sometimes, however, the past gets in the way. We hold on to things that happened to us, which hold us back from finding our full potential. We don't realize that we need to shift from how we operated in combat or in the military to a different way of operating in our post-military life. That's where talking to a mental health professional can be beneficial: setting down the sword and shield and learning how to live in a future of peace.

A Future of Productivity

Continuing on the idea of our post-military lives: getting stuck in the past can hamper our productivity in the future. If the weight of the past is binding us, then we are not as productive as we can be. If we are struggling with depression, or isolation, or guilt, or any number of psychological challenges, then we're not going to get as much done as if we didn't have to deal with that stuff. Sleepless nights lead to crappy days, and anger and frustration lead to misjudged relationships.

I often describe to my veteran clients that removing the negative does not automatically lead to increasing the positive. Just because we stop being sad doesn't mean that we are going to be happy; it's not just about getting back up to zero (a lack of a painful emotion) but about getting *above* zero, to a life of satisfaction. By becoming aware of why we do the things we do, and then deciding to change them if we want to, we can get to life of productivity in spite of what happened to us or what we did.

A Future of Acceptance

As a friend and colleague often says, "You will never be a civilian again." That's the truth. Anyone who has served a significant time in the military is changed from what they were before. We are no longer service members, Soldiers, Marines, Airmen, etc. But we are also not a "civilian." We're this entirely third thing, a "veteran," with all of the benefits and detriments that go along with it. We see the world differently. We think more globally than someone who has never left their state, much less the country. Many of us, those who have been to combat, have seen the worst that humanity has to offer. Coming to terms with that can be difficult, but not impossible; working with a mental health professional can help make that change.

Working Now for a Pain-Free Future

I see if often in veterans I work with. After an understanding of what they experienced, recognizing how they've coped with it in the past hasn't gotten them the outcomes that they wanted, and learning ways to get different outcomes, they get to a place of *peace*. Of acceptance. I tell them, "I'm not concerned with right now, or next week or next month. I want to help the you thirty years from now." That's what I'm focused on; how do I help the veteran in front of me find peace in the future, so that when their grandkids say, "Grandma, Grandpa, what did you do in the war?," they can tell them about their experiences without taking a shot of whisky or breaking down in tears.

That's what working with a therapist can do; head things off at the pass, set the psychological bone right, so to speak, so that you can have a pain-free and productive future.

Veterans, Do You Strive for Success or Significance

When I talk about the work that I do with veterans as a mental health counselor, people are often surprised when I tell them that we don't talk about war all the time. Every conversation isn't about rehashing old memories, tearing off old scabs, and opening duffel bags that have been locked in a closet for years. Many times, we're talking about what to do next, what to do in their post-military lives: how do I become successful? Instead, however, I take things in a different direction…instead of trying to achieve success, what about trying to achieve significance?

I was listening to a conversation between two entrepreneurs one day[11], and one of them talked about doing something meaningful in your life. He talked about this quote that he saw written on a wall in Philadelphia:

You May Be Successful, But Will You Matter?

That got me thinking: when a veteran leaves the military, do they know what they're working towards? A house with a white picket fence? When we leave the military, whether it's after six years or twenty-six years, there's the question: what next? What do I want to *do,* what do I want to *be*? For may veterans, that's the most important question that is never asked.

Success is defined in several ways, but Webster defines it as the *accomplishment of an aim or purpose,* the *attainment of popularity or profit,* or *a person or thing that achieves desired aims or achieves popularity.* That means that our value is tied to something external to us: whether people like us or whether we have a lot of money. Significance, on the other hand, is defined as *the quality of being worthy of attention; importance.* This is an INTERNAL value…even if no one pays attention to me, am I *worthy* of attention? Deciding whether we are going to strive for success or significance is a choice that make a big difference at the end of the journey.

Significance Doesn't Mean Being Well Known

There are a lot of reasons people may want to be well known. It may be ego, "I can only measure my worth by my position against those around me". It may be that you want to be well known in order to be able to help that many more people. The challenge is, if being well known is your goal, then any way you get there is acceptable. You will start to do things to be noticed, and it may not be in a good way…but hey, at least you're well known. Infamous is as good as famous.

When I was a teenager, I went to live on my aunt and uncle's farm in rural Missouri for a summer. Not a Silence of the Lambs thing, just a summer hanging out on the farm. I had been there for two and a half months when we went into town, and this good ol' boy at the local diner turns to me and my cousin and says, "You're the new boys living at Ron and Vicki's, right?" Everybody knew us, because *nobody* new ever came around. We weren't significant, though; we didn't matter.

Significance Doesn't Mean Being Rich

In the same way that being well known isn't all that it's meant to be, being "rich" is not a requirement for significance. Being wealthy or having a great number of assets doesn't mean that you make a difference in the world. Having a positive impact on others is the critical thing, and that doesn't always have to result in personal enrichment. After leaving the military, getting a job that pays the bills is great, and having less bills to pay is always a good thing. Once that's in place, though, then what?

I had a veteran ask me the other day, "if you woke up tomorrow and had no obligations, no responsibilities, what would you do?" I considered that for a bit. If I had enough money where I didn't have to do anything, everything was taken care of, what would happen? Personally, I would create obligations and responsibilities for myself. I'd find someone to talk to, probably about veteran mental health. I would find a problem to solve (or create a problem that needed solving) just so I had something to *do*. So having money is a measure of success, not significance.

Significance Doesn't Mean Making a Huge Impact in the World

So if we're not striving to be well-known or rich after the military, what are we working towards? Making a difference in the world? What does that even mean? We look at the mountaintop from the plains and say, "if only I could get there, I could make a difference." How often do we miss an opportunity to be meaningful or significant on the way to the top, though? We don't have to be significant to the world, we can be significant to our world.

I've written before about how veterans of today have the ability, and the responsibility, to make an impact on this century the way that the post-WWII generation made an impact on the last century. Yes, they became scientists and lawyers and presidents and CEOs. They also became, like my grandfather, tailors and mechanics. Veterans who became significant in their own way, in their own lives. My mother has fifteen brothers and sisters, with one more adopted. While my grandfather wasn't famous, and he sure wasn't rich, he was certainly significant! He was important to us, our family, our community.

Significance Means Making an Impact Where You Are, How You Can

So make an impact. Do something, or work towards something, that is greater than yourself. It's what we did when we were in the military, it's what we can continue to do after we're out. It's the key part of the purpose and meaning stuff we always talk about. What are you going to do today to be significant? What is the impact that you're going to have on your world, in a small way, that is going to make all the difference? Let me know in the comments section, so we can celebrate your significance.

Are You Willing to do Anything to Overcome Obstacles?

I have some questions for transitioning veterans. Are you a driven and successful professional? Will you do anything to overcome obstacles on the path to success? When I was in the military, I had a leader who said that his motto was, "Accomplish the mission by any means necessary." I mentioned to him, of course, that he meant "by any legal, moral, and ethical means necessary," and he rolled his eyes. But I knew what he meant. Barring anything unethical and illegal, though, you would not stop at anything to reach your goal, right?

This is something veterans know about. We may not know interviewing, KSAs, or be familiar with networking, but if there's one thing we DO know, it's getting stuff done. Early morning, late night, weekends if necessary, our ability to put up with immense amounts of stress in order to get the job completed is often a point of pride.

So if we are willing to do literally anything to succeed in our mission of transitioning out of the military and find a meaningful and fulfilling life, why are we letting little things get in our way? Those habits that we have, the mindset that we bring to the table, are often the biggest obstacles to our success. Here are three questions that may be counterintuitive to action people, but may be more effective than bashing our heads against a brick wall.

Are You Willing To Focus On The Needs Of Others Instead Of Your Own?

I've heard a number of variations on this theme. "Show them how you are the solution to their problem" and "Identify a potential employer's pain then explain how you can solve it." That's great…but it starts with listening to *them*, or doing research about what challenges they have. More importantly, it is about listening to another *person*, as a person and not as their role. Not the HR guy, or the owner of the small business, or anything else. Figuring out what challenges that they have…that's it. It's not like you're a cat sitting outside of a mouse hole, and the minute you find something that you do well that you then pounce on it and say, "Aha! I can help you there!" This is about resisting the natural inclination to satisfy your own needs, and instead focus on the needs of others. This approach…serving, instead of being served…can bring about the results you desire. It's not manipulating, it's not being disingenuous, it's about truly caring about what that other person's needs are. Shining the spotlight on them well and they may eventually say, "man, you sure handle that spotlight well. What can I do for you?" Even if they don't do that immediately, never underestimate the personal satisfaction value of doing good service to others.

Are you willing to let go of your hangups in order to see the bigger picture?

You found a job description that is perfect for you. The only thing missing from it is your name. The research is done, you've networked, you've tailored your resume, you've done everything you could and have been taught to do in order to get your foot in the door. You submit your resume, send follow-up emails and even call a couple times…and nothing. You're waiting for the phone to ring. A natural inclination is to allow frustration to take over, possibly even anger. Frustration and anger often have a target, and the target can be that company. Or that hiring manager. You start talking to your friends and family: "man, you won't believe this. They haven't called me back! How can they not know that I'm perfect for this job?" Your friends and family, caring about you as they do, jump right on board with you and help move the train down the tracks. "I know, right? They don't know what they're missing, they'd be stupid not to hire you!" they say. "They are stupid, aren't they? They're a bunch of…." At this point, though, I'm making it about me. I have no clue what's going on inside the company. And how many other opportunities have I overlooked, fixating on this one "perfect" opportunity? This is just one possible path to your ultimate destination, success and a satisfying life. Standing there waiting for a gate to open doesn't get you any further down the path.

Are you willing to stop focusing on what should be instead of what is?

Sometimes, it's the imperatives that get in our way. Things *should* be a certain way, they *must* happen a certain way, they *have to* be how I think they *should* be and I *must* do all that I can to make it happen. How often do we let the shoulds and musts get in the way of reality? Many argue that veterans have a sense of entitlement, and some veterans actually do have that. "I'm a veteran, I fought for this country, that should count for something!" Maybe it does to some people, maybe it doesn't to others. Thinking that it *should* count for something does not change the fact that it might not count for something. Denying reality is a great way to fill your mind, and your day, with frustration. Once we start accepting the reality as it is instead of rejecting the reality in favor of what we think it should be, we can be much more flexible in our approach to our goal.

There are certainly obstacles to us obtaining our goal. Any goal worthy of effort will have obstacles, and some may argue that it is the struggle that makes the goal worthy. There are many ways to overcome obstacles, however, and not all of them are through direct action. Perhaps some obstacles are overcome by avoiding them, or by stepping away from them.

If you're spending your time trying to shove a stubborn mule out of your path, the only thing that may happen is that you get tired shoulders and a pissed off mule. Neither of you are better off for your efforts. If you are willing to do anything to succeed, is that entirely true? Does that anything include stepping outside of your comfort zone and focusing on others for a while? Taking the time to focus on others, letting go of your hangups, and accepting reality will make for a much more pleasant journey, and you'll reach your goal before you know it.

Acknowledgements

A sincere and grateful thanks to everyone that helped me become who I am. My parents, Mary and Duane, instilled in me discipline, respect, care for others, and a servant's heart. While military service was not their first choice for me, they supported me all those years ago when I made the choice to join. My wife, Connie, thought knew what she was getting into when she joined me at Fort Bragg nearly twenty years ago to start our life together, and has remained a constant support for me through all the twists, turns, ups, and downs that have come along since. My children, Christina and Daniel. The pride and love that I have for you is greater than you may ever know, and it is my sincere hope that you will live lives of joy and peace, never experiencing the sounds of war. To my sisters, Tammy and Jennifer, and my brothers, Brett and Jesse, and my aunts, uncles, nephews, and cousins (way too many to mention); though life, time, and distance has spread us out across the nation and the world, your love and support for me is a constant reminder that we're never truly alone. For all of those I served with, from Germany to Bragg to Germany to Recruiting Command to Fort Carson: a huge number that I lived, loved and fought alongside (and sometimes with). Any amount of success that I had as a Soldier and a Leader is directly due to the brothers and sisters around me. To my mentors in my second life as a mental health professional: Susan Varheley, Jared Theimann, your instruction and guidance at the beginning of my career were invaluable, and made me proud to be an Adams State University alum. For Jenni Guentcheva and Donya Boudeman, in taking a chance on a bold combat vet who had no idea what he was getting himself into when he said he wanted to help other combat vets; your supervision and guidance were and is still invaluable. For Drs. Chuck and Rae Ann Weber, and all the team at the Family Care Center, your support for me is nothing compared to your unwavering support for our service members, veterans, and their family. To the National Board of Certified Counselors Foundation, especially Sherry Allen and (of course) Drs. Dannette Patterson and Greg Frazier; again, taking a chance on me and providing the support needed to get out of my comfort zone. I am truly standing on the shoulders of giants. To Judge David Shakes of the Colorado Fourth Judicial District Veteran Trauma Court, and the rest of the VTC Team: the dedication you provide to our justice involved veterans is inspirational, and I'm proud to work with you. To retired Command Sergeant Major Dan Elder, your continued support and encouragement helped the Head Space and Timing blog get off the ground, and none of any of this would have been possible without your guidance and assistance. And finally, and certainly not least, to the family and memory of Sergeant Eduviges Guadalupe Wolf. Duvi, your sacrifice on that terrible October day will always serve as a reminder of the reason why I do what I do. Rest In Peace.

Notes

1: Raising Awareness About The Psychological Impact Of Military Service

1. Frequently Asked Questions regarding participation of licensed professional counselors in the Veterans Affairs (VA) health care system. American Counseling Association Office of Government Affairs: www.counseling.org/government-affairs/public-policy)
2. Medicare and Professional Counselors, National Board of Certified Counselors, http://www.nbcc.org/govtaffairs/medicare)
3. ASIST two-day training. ASIST is an interactive two-day workshop that provides participants with an understanding of suicide intervention techniques. https://www.livingworks.net/programs/asist/
4. What is CTE, the Concussion Legacy Foundation, https://concussionfoundation.org/CTE-resources/what-is-CTE
5. The Terrorist inside my husband's brain, September 2016, Susan Schneider Williams. Editorial in the journal Neurology, http://n.neurology.org/content/87/13/1308.full
6. A Father's search for meaning takes him to battlefield where son died. Jeremy Schwartz, May 2012, Austin Statesman
7. Dreazen, Yochi. *The invisible front: Love and loss in an era of endless war.* Broadway Books, 2015.
8. VA Secretary announces intention to expand mental health care to former service members with other-than-honorable Discharges and in Crisis. VA Press Release, March 2017
9. Ready to Serve: Community-Based Provider Capacity to Deliver Culturally Competent, Quality Mental Health Care to Veterans and Their Families. RAND corporation report, Tanielian, T., Farris, C., Epley, C., Farmer, C. M., Robinson, E., Engel, C. C., & Jaycox, L. H. 2014
10. Cognitive Impairment and the Neurological Basis for PTSD. Dr. J. Blair Cano, guest post on the Head Space and Timing blog, July 2017
11. Trauma & Stressor-Related Disorders, Posttraumatic Stress Disorder, Diagnostic and Statistical Manual of Mental Disorders (DSM-5), 2013
12. The Army response to Hurricane Katrina, Roberta Berthelot, September 2010. https://www.army.mil/article/45029/the_army_response_to_hurricane_katrina

2: Developing Personal Awareness of the Need for Veteran Mental Health

1. Cyril Richard Rescorla, or "Rick," was an iconic veteran of British origin who fought with 1/7 Cav during the Battle of Ia Drang, immortalized in Hal Moore and Joseph Galloway's book. Rick was a security officer for Morgan Stanley Dean Witter on September 11th, 2001, and is credited for saving numerous lives during the attack on the World Trade Center. For more information: www.rickrescorla.com
2. Blog Post written October of 2016, "The Moment I Realized I Hate War as Only a Warrior Can, http://veteranmentalhealth.com/2016/10/18/the-moment-i-realized-i-hate-war-as-only-a-warrior-can/
3. For more information about substance abuse and addictive disorders, see the DSM 5, American Psychiatric Association. (2013). *Diagnostic and statistical manual of mental disorders* (5th ed.). Arlington, VA: American Psychiatric Publishing

3: Developing Resilience to Recover

1. Willner, D. (1982). The Oedipus Complex, Antigone, and Electra: the woman as hero and victim. *American Anthropologist, 84*(1), 58-78.
2. Maslow, A., & Lewis, K. J. (1987). Maslow's hierarchy of needs. *Salenger Incorporated, 14,* 987.
3. Bracha, H. S., Ralston, T. C., Matsukawa, J. M., Williams, A. E., & Bracha, A. S. (2004). Does "fight or flight" need updating?. *Psychosomatics, 45*(5), 448-449.
4. Frankl, V. E. (1985). *Man's search for meaning*. Simon and Schuster.
5. A discussion of the weird stuff veterans pick up from their time in the service; I have a worn-out baseball on my desk that has more meaning to it than just a baseball. http://veteranmentalhealth.com/2016/08/23/the-baseball-on-my-desk
6. Daily Dose with Bennett and Eddie, Episode 68, "War Movies Are My Penance" released 11 October 2017
7. Kipling, R., & Eliot, T. S. (1943). *A Choice of Kipling's Verse*. Macmillan of Canada
8. MDMP and TLP are part of the Army's Tactical Decision Making Process. For more information, look up Army Field Manual 3-21.8, Infantry Rifle Platoon and Squad
9. Linehan, M. (2014). *DBT® skills training manual*. Guilford Publications.
10. Reivich, K. J., Seligman, M. E., & McBride, S. (2011). Master resilience training in the US Army. *American Psychologist, 66*(1), 25

4: Developing Skills to Apply to Our Post-Military Life

1. Abramson, L. Y., Seligman, M. E., & Teasdale, J. D. (1978). Learned helplessness in humans: Critique and reformulation. *Journal of abnormal psychology, 87*(1), 49. For a further discussion about learned helplessness and veteran mental health, see http://veteranmentalhealth.com/2016/08/06/helping-veterans-trapped-by-their-own-experiences-learned-helplessness-and-veteran-mental-health
2. Linehan, M. M. (2014). *DBT? Skills Training Handouts and Worksheets*. Guilford Publications
3. Linehan, M. (2014). *DBT® skills training manual*. Guilford Publications

5: Personal Satisfaction in our Post-Military Life

1. I found an excellent article that describes Jung and the Wounded Healer, and the origins of the concept, which you can read here: http://www.jungatlanta.com/articles/fall12-wounded-healer.pdf
2. Stacy is a fellow clinical mental health professional and host of the Speaking of Suicide blog. Her article can be found at https://www.nytimes.com/2017/05/11/well/mind/a-suicide-therapists-secret-past.html
3. A reflection on how painful emotions, such as PTSD and Depression, can be as comfortable as a faithful black dog. http://veteranmentalhealth.com/2017/02/07/the-black-dog-of-the-veteran-emotion
4. A discussion about how the many challenging thoughts in a veteran's mind can be like Pandora's Box. http://veteranmentalhealth.com/2017/01/19/the-pandoras-box-of-the-veteran-mind
5. White, S. (2000). *Critical conditions*. Hampton Falls, NH: Beeler

6: The Importance of Remembrance Post-Military Life

1. Army Techniques Publication (ATP) 1-05.02 establishes a common understanding, foundational concepts and methods for executing religious support (RS) during funeral services and memorial ceremonies and services. Find more at https://armypubs.army.mil/epubs/DR_pubs/DR_a/pdf/web/atp1_05x02.pdf

2. From the Marine Corps Club and Hotel Website: The flagship Marines' Memorial Club is located in the heart of downtown San Francisco, just off Union Square and the cable cars. This handsome California Spanish Revival hotel offers an inviting club atmosphere that is rich with history, honor and pride. Find more at https://marinesmemorial.org

3. 2LT JP Blecksmith was a Marine Corps officer assigned to 3rd Battalion, 5th Marine Regiment, 1st Marine Division, I Marine Expeditionary Force, Camp Pendleton, Calif. The Marine died supporting Operation Iraqi Freedom; he was 24. Blecksmith was an athlete and a graduate of the United States Naval Academy. He was survived by his parents, Edward and Pamela Blecksmith, his brother, and sister. You can learn more about the service and sacrifice of Lieutenant Blecksmith at https://www.travismanion.org/fallen-heroes/2nd-lt-james-p-blecksmith-usmc

7: Applying Lessons Learned to Post-Military Life

1. In a Clock Time culture, individuals let an external clock determine when tasks will occur; in an Event Time culture, tasks are planned in their relative position to other events. Find more at Avnet, T., & Sellier, A. L. (2011). Clock time vs. event time: Temporal culture or self-regulation?. *Journal of Experimental Social Psychology*, 47(3), 665-667

2. Csikszentmihalyi, M. (1997). Flow and the psychology of discovery and invention. *HarperPerennial, New York*, 39

3. The After Action Review helps provide feedback on mission and tasks in training and combat. It helps to identify how to correct deficiencies, sustain strengths, and focus on specific tasks. Find more at Army Training Circular 25-20, A Leader's Guide to After-Action Reviews https://www.acq.osd.mil/dpap/ccap/cc/jcchb/Files/Topical/After_Action_Report/resources/tc25-20.pdf

8: Veteran Mental Health as Mental Wellness instead of Mental Illness

1. This study, conducted by the National Academies of Sciences, Engineering and Medicine, interviewed over 4,000 veterans nationwide about their mental health needs. Find out more at http://nationalacademies.org/hmd/Reports/2018/evaluation-of-the-va-mental-health-services.aspx

2. Dr. Greg Frazier used this analogy during our conversation on episode 11 of the Head Space and Timing Podcast. Check out our conversation by going here: http://veteranmentalhealth.com/2017/09/12/hst-011-national-board-certified-counselors-foundation-veterans-initiatives-greg-frazier

3. Dr. Seligman has been a leading voice in the field of positive psychology for over forty years. Check out the work that he and his team are doing at https://ppc.sas.upenn.ed

4. Seligman, M. E. (2011). Flourish: a visionary new understanding of happiness and well-being. *Policy*, 27(3), 60-1

5. Peterson, C., & Park, N. (2005). Positive psychology progress. *American Psychologist*, 60(5), 410-421

6. Mental Health America is a national community-based nonprofit that is dedicated to addressing the needs of those living with mental health conditions that impact their lives. Find out more about the Before Stage 4 campaign at http://www.mentalhealthamerica.net/b4stage4-philosophy

7. Posttraumatic Growth is the idea that we can emerge out of stressful and traumatic events stronger than we were before we experienced them. For further reading on Posttraumatic Growth, check out "Through the Other Side of the Valley of Death: Veterans and Posttraumatic Growth" on the Head Space and Timing blog

8. Csikszentmihalyi, M. (1990). Flow: The psychology of optimal experience. New York: Harper Perennial

9. The Warrior Ethos is four lines in the Soldier's Creed, which embody the mindset that a warrior should have. Find out more at https://www.army.mil/values/soldiers.html

10. Dannette Patterson is a military spouse and clinical mental health professional that firmly believes in the power of mentorship. Find out more about her and her work on Episode 40 of the Head Space and Timing Podcast.

11. This discussion occurred on the Legends and Losers podcast, between host Christopher Lochhead and his guest, Jeff Hoffman. If you are looking for self-improvement, you can do much worse than listening to this podcast: www.legendsandlosers.com

ABOUT THE AUTHOR

Duane France is a retired U.S. Army Noncommissioned Officer, combat veteran, and clinical mental health counselor practicing in Colorado Springs, Colorado. In addition to his clinical work, he also writes and speaks about veteran mental health to a wide audience through his podcast and blog, Head Space and Timing, which can be found at www.veteranmentalhealth.com

Made in the USA
Middletown, DE
11 December 2019